Energy Innovations
2017

Also by Emil Morhardt

Energy Innovations
2017

J. Emil Morhardt, Editor

CloudRipper Press

Cutting Edge Books

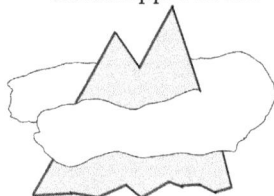

CloudRipper Press
Santa Barbara, California
www.CloudRipperPress.com
Copyright © 2017 J. Emil Morhardt
Morhardt, J. Emil
Energy Innovations, 2017
J. Emil Morhardt, Editor.
ISBN 978-0-9963536-8-7 (paper)

TABLE OF CONTENTS

Vehicles 271

FORWARD

by Emil Morhardt

Energy seems, on the one hand, to be the most entrenched of commodities—almost all of our electricity comes from large legacy coal- or gas-fired power plants connected to the grid which is, in turn, seamlessly connected to every wall socket in our homes and businesses; and almost all of our vehicle fuel comes from legacy oil refineries, delivered to service stations in tank trucks. It has been this way for my entire lifetime and probably was for most of our parents and even grandparents' lifetimes as well, and it has hardly changed at all. As a high school student I pumped gas in a Texaco station for several years, equipping me well for the self-service pumps we all use today; the pumps have not visibly changed. As a young man I worked as a consultant at many power plants and oil refineries—they haven't visibly changed either.

On the other hand I drive a hybrid car stuffed with new technology that I can't even begin to understand, and don't need to stop at gas stations as often as before, and I often drive it through the expansive photovoltaic generating stations, and wind turbine farms of the Antelope Valley, the solar thermal power towers in adjacent Lancaster, and the solar thermal troughs in Kramer Junction, all in the Mojave Desert, on my way up to the Owens valley where I have a house along a creek between two of the many hydroelectric power plants fed by the runoff from the Sierra Nevada Mountains. There is clearly a change in our energy mix in the offing, and global warming, or the fear of it, is driving many of these changes, and the many more that are coming.

Because it seems likely that there is money to be made by reducing fossil fuel use, and energy use in general, there is currently an intense amount of entrepreneurial activity surrounding all aspects of our energy supply and usage. This book is an attempt to make some sense of the overwhelming amount of information about this activity streaming down the web (and in the scientific and engineering journals). It is a result of three months in early 2016 of combing through entrepreneurial websites, news items in the press, and a variety of other sources (all well documented in the book) to see what new and exciting developments are occurring in energy. The result is a fascinating look at the types of changes in our energy mix in the near future.

The text consists of more than 250 vignettes of innovative energy projects, many in their earliest stages, organized by type of energy being considered, then by author.

Because this book was written during politically tumultuous times—President Trump was inaugurated in the week after the class started—many of the summaries discuss things in progress at the time, and so might seem dated. But the book was published 5 months into the Trump first term, so they are still fresh in the authors' minds.

RENEWABLE ENERGY

Photovoltaics

Samoan Island Converts to 100% Solar

by Aurora Brachman

The island of Ta'u, an outer island of American Samoa, is now entirely run on solar energy. This was made possible by the creation of a new solar-powered microgrid that transitioned the island from 100 % diesel fuel to 100% solar. The microgrid was implemented by SolarCity, a California-based company that was recently bought by Tesla's Elon Musk. The Project was funded by the U.S. Department of the Interior and the American Samoa Power Authority and cost a total $8 million to complete. The system includes 5,328 solar panels and generates 1.410 megawatts of electricity. The energy is stored in 60 Tesla Powerpacks and can be stored for up to three days without sunlight. Given the extreme weather conditions on the island, they are designed to be capable of withstanding category 5 hurricane winds. Many other islands are working to achieve energy independence through solar energy but Ta'u is the first to successfully do so.

Despite some concern, life on the island has remained essentially the same since making the switch from diesel to solar. When SolarCity officially "flipped the switch" to turn on the solar power plant and disable the diesel generators, the lights on Ta'u hardly flickered. SolarCity credits the project's success partly to the seamless transition that can be made from the old power source to solar.

The success of this project is a promising sign for the future of sustainable energy for many island nations. This proves renewable energy can be reliable, well received by indigenous populations, and ultimately a worthwhile investment. The elders of Ta'u see this shift to renewables as a return to their distant past when they were able to be more self-sufficient and could live harmoniously with their environment.

Lin, Daniel. "How a Pacific Island Changed From Diesel to 100% Solar Power." National Geographic. National Geographic Society, 23 Feb. 2017. Web. 25 Feb. 2017. (http://news.nationalgeographic.com/2017/02/tau-american-samoa-solar-power-microgrid-tesla-solarcity/)

Byrne, Kevin . "Solar energy powers an entire island in American Samoa thanks to Tesla, SolarCity." Local Weather from AccuWeather.com - Superior Accuracy™. AccuWeather , 12 Jan. 2017. Web. 25 Feb. 2017. (http://www.accuweather.com/en/weather-news/solar-energy-powers-an-entire-island-in-american-samoa-thanks-to-tesla-solarcity/70000524)

Call to Action: Urge to Take Advantage of Solar Energy Potential in Mexico City Metropolitan Area

by Alejandra Chávez

The Mexico City Metropolitan Area (MCMA) has a population of 8.851 million people (2010), making it the largest and most populated urban area in the country. It is well within the "sun-belt" of Earth, with solar incidence areas of over 5 kWh/m²/day radiation. Though there is a high potential for renewable energy, there is little incentive by policymakers to take advantage of an energy source that could better conserve the environment, limit the amount of natural and/or technical interruptions, and be an economic relief to the MCMA. The authors support the development of and rationalize that solar photovoltaic (PV) technology can harvest its energy potential. This technology is economically competitive to give energy to vulnerable areas that often do not require as much power. Because energy transition diagnostic is not very organized, it is vital for decision-makers to be aware of the social and economic difficulties in MCMA.

Within the MCMA, there about 63% of the population has a 'medium' to a 'very high' degree of marginalization. In this text, marginalization is defined by the National Population Council, which explains that is it when a community is at a disadvantage via exposure to hazards and instabilities. More specifically, these communities in MCMA have significant restrictions when adapting to energy conditions; thus, they develop a higher dependency on the resources that surround them. For these reasons, the authors call on the State to ensure that the governmental policies that focus on energy issues and social exclusion be enforced. Ultimately, the authors question if the local and federal public strategies are willing to boost social development and urban sustainability. Their research focuses on analyzing the social and environmental agenda to further investigate interests between the Mexico City Metropolitan Area, techno-environmental marginalization, and solar photovoltaic (PV) technology-based energy transition.

Arenas-Aquino, A.R., Matsumoto-Kuwabara, Y. & Kleiche-Dray, M. Environ Sci Pollut Res (2017). doi:10.1007/s11356-017-8387-9
Springer Link (http://link.springer.com/article/10.1007/s11356-017-8387-9)

First Trafficable Solar Road in the U.S. Arrives in Georgia

by Alejandra Chávez

In December of 2016, the first trafficable solar roadway in the United States was installed on an 18-mile stretch of Interstate 85 (I-85) in West Point, Georgia. This section of the Interstate was renamed the Ray C. Anderson Memorial Highway, named after a Georgia native who is recognized as a leader in green business, and is commonly known as

The Ray. The Ray uses Wattway solar panels, which first debuted in France in 2015, to generate energy from the sun when not blocked by vehicles. The innovative drive-over solar panel technology consists of thin, heavy-duty, and skid resistant photovoltaic pavers that can be applied directly over existing asphalt. The need to remodel or build new road infrastructures is thus eliminated.

The clean energy generated by the solar road will help power the Georgia Visitor Information Center, with the potential to power public lighting and road signals in cities, towns, and remote areas where low population density increases the cost of connecting the grid. This technology could also be used to provide an alternative energy solution for bike paths, shopping centers, airports, electric vehicle charging stations, and other infrastructures Allie Kelly, the executive director of The Ray, believes that installations of solar roads can become a widespread energy generation system for the department of transportation and eventually become profitable.

Highways are among the most polluted areas—five million tons of CO_2 are emitted every year just from this 18-mile stretch of road. The Ray is combating that pollution by becoming "the site of a highway solar farm" that can produce one megawatt of energy. The Georgia Department of Transportation has taken on projects to funnel the energy generated from The Ray to an electric vehicle charging station and power tire pressure monitoring technology near the Georgia-Alabama border.

[http://theray.org/ 2016/12/20/wattway-debuts-on-the-ray/].
Sisson, Patrick. "In rural Georgia, tomorrow's smart, sustainable, solar highway is being built today." Curbed. Vox Media, 6 Feb. 2017. Web. 11 Feb. 2017.
Curbed (http://www.curbed.com/2017/2/6/14521102/highway-the-ray-solar-power-transportation)
Lanier, John A. "The Ray Unveils Two New Transportation Technologies – Both Firsts in the U.S." Ecocentricity Blog. The Ray C. Anderson Foundation, 19 Dec. 2016. Web. 11 Feb. 2017.
The Ray C. Anderson Foundation (http://www.raycandersonfoundation.org/articles/the-ray-unveils-two-new-transportation-technologies-both-firsts-in-the-us)
"Wattway Debuts on the Ray." The Ray C. Anderson Foundation. The Ray, 20 Dec. 2016. Web. 11 Feb. 2017.
The Ray (http://theray.org/2016/12/20/wattway-debuts-on-the-ray/)

Pocket-sized Solar Lamps 'Bring Light' to People in Remote Locations

by Alejandra Chávez

Olafur Eliasson, a Danish-Icelandic artist, revealed his third design of mini solar-powered lamps at the 2017 Design Indaba Conference in Cape Town, South Africa. The first design, The Little Sun Charge, launched in 2012 and was finally realized through a Kickstarter crowdfunding campaign in 2015. The third design, known as the Little Sun Charge, has been improved to be more efficient; it can produce five hours of light when charged in the sun for five hours. The lamp can alternate between a "concentrated reading-style light" and a "magical, sparkling glow".

The lamp was designed for people who do not have access to reliable energy. The founder's hope is to "bring light" to people that are in remote locations in the hopes of reducing their reliance on kerosene lanterns, which have proven to be costly, inefficient, environmentally unsafe, and whose fumes have extensive health drawbacks. By using the Little Sun lamps, people can save on the cost of fuel needed for the lanterns and invest them elsewhere in the country. Eliasson even claims that because of the Little Sun project, $55 million were freed up things other than petroleum.

Though these lamps are intended for communities without reliable electricity, anyone is encouraged to purchase the Little Sun lamps because they can help lower the cost of the product in Zimbabwe, Ethiopia, Kenya, Senegal, Ghana, among others. At the 2017 Design Indaba Conference, it was announced that over 220,000 lamps have been sold in Europe and the United States—a total of 280,000 lamps worldwide. In the future, the company hopes to continue its efforts to fund their project with different non-governmental organizations and private sector collaborations. More importantly, the Little Sun lamps intend to "raise awareness, raise trust, and raise the belief that solar is a viable option."

http://www.curbed.com/2017/3/7/14843110/olafur-eliasson-little-sun-diamond-lamps].
http://www.curbed.com/2017/3/7/14843110/olafur-eliasson-little-sun-diamond-lamps].
Aouf, Rima S. "Olafur Eliasson reveals his latest pocket-sized solar lamp, the Little Sun Diamondt." Dezeen. Dezeen, 7 Mar. 2017. Web. 15 Mar. 2017.
DeZeen (https://www.dezeen.com/2017/03/07/olafur-eliasson-design-new-pocket-sized-solar-lamp-little-sun-diamond/)
Cheah, Selina. "Olafur Eliasson's 'Little Sun Diamond' lamps bring light to remote locations." Curbed. Vox Media, 7 Mar. 2017. Web. 15 Mar. 2017.
Curbed (http://www.curbed.com/2017/3/7/14843110/olafur-eliasson-little-sun-diamond-lamps)

Apple Named the Cleanest Company in the World

by Alejandra Chávez

According to Greenpeace's 2017 renewal energy report, "Clicking Clean: Who is Winning the Race to Build a Green Internet," Apple is the greenest tech company in the world for the third year in a row. In the fiscal year of 2016, Apple's Clean Energy Index was 86%, surpassing Facebook by 16%, Google by 27%, and HP by 33%. The company's comprehensive carbon footprint was 29.5 million metric tons, an impressive 23% decline from 2015. Of all the major data center operators, Apple was the only tech company to use less than 5% natural gas, coal, and nuclear power plants as energy sources. The Greenpeace report also mentions that Apple made the commitment to be 100% renewably powered in 2012, something the company has restated in its 2017 Environmental Sustainability progress report — "We mapped our carbon footprint, and we're working to eliminate it." To transition completely to renewables, Apple plans to target five major areas: manufacturing, product use, facilities, transportation, and recycling. In terms of facilities, Apple's new corporate headquarters in Cupertino, California follows this model because it will run completely on renewable energy, with about 700,000 square feet of solar panels. In terms of recycling, the company mentions that 99% of the paper used for product packaging is from recycled materials.

Although Apple currently has 485 megawatts of wind and solar projects in China alone, the company hopes to generate and source from more than four gigawatts of new clean energy worldwide by 2020—enough energy to power four million homes. Apple's efforts to transition to zero carbon emission does not stop at its company; according to the Business Insider, the company actively pushes other IT data center and cloud operators to "follow their lead" in powering their operations with renewable energy.

"Clicking Clean: Who is Winning the Race to Build a Green Internet?" Greenpeace. Greenpeace Inc., Jan. 2017. PDF. 23 Apr. 2017. https://www.google.com/url?sa=t&rct=j&q=&esrc=s&source=web&cd=1&cad=rja&uact=8&ved=0ahUKEwiUvtjZ8L7TAhWIslQKHQBbBGQQFggkMAA&url=https%3A%2F%2Fwww.greenpeace.de%2Fsites%2Fwww.greenpeace.de%2Ffiles%2Fpublications%2F20170110_greenpeace_clicking_clean.pdf&usg=AFQjCNHLEAIy0QoTTlBbt890Rl8Sxb9-6Q&sig2=_WdsV4tuHsuNxgBDjVlSdQ

"Environmental Responsibility Report." 2017 Progress Report, Covering FY2016. Apple, 2017. PDF file. 23 Apr. 2017. https://images.apple.com/environment/pdf/Apple_Environmental_Responsibility_Report_2017.pdf

Varinsky, Dana. "Apple is the greenest tech company in the world, according to Greenpeace." Tech Insider. Business Insider Inc., 10 Jan. 2017. Web. 23 Abr. 2017. http://www.businessinsider.com/apple-greenest-tech-company-greenpeace-2017-1

Wattles, Jackie. "Where Apple stands in its quest for clean energy." CNN Tech. Cable News Network, 23 Apr. 2017. Web. 23 Apr. 2017. http://money.cnn.com/2017/04/22/technology/apple-clean-energy/

Acid Technology

by Dominique Curtis

Soon there could be a pill that takes measurements of your body temperature and heartbeat and sends them straight to an app on your phone. Researchers have discovered a new system to sense vitals and possibly deliver drugs to patients. Researchers at MIT and Brigham and Women's Hospital have designed and began testing a small voltaic cell that is sustained by the acidic fluid in the stomach. Researchers Traverso and Langer have been working on building and testing many ingestible devices that can sense physiological conditions such as temperature, heart rate, and breathing rate. They also think that these devices might be able to deliver drugs to treat diseases such as Malaria. They started by taking the temperature of pigs using the ingestible device, and sending the information to a receiver two meters away.

The device is a 12 x 50 mm cylinder and researchers predict that it will be one-third that size in the future. It is made by attaching zinc and copper electrodes to its surface; of the zinc emits ions into the acid in the stomach to power a voltaic circuit. The circuit generates energy to power a commercial temperature sensor and a 900 megahertz transmitter (MIT Researchers, 2017).

Many researchers in the field of medicine and engineering are enthusiastic about the next stages of these devices. Biomedical engineer Chandrakasan says that this can lead to "new medical devices where the body itself contributes to energy generation enabling a fully self-sustaining system." Other researchers are interested in how to use this device to try out different dosages of drugs. They are working on a pill that releases drugs that are encapsulated by a gold film in the voltaic cell. One of the challenges researchers admit they have to overcome before they can think bigger is monitoring and managing energy generation, conversion, storage, and utilization.

Massachusetts Institute of Technology. "Science News." ScienceDaily. ScienceDaily, 6 Feb. 2017. Web. 12 Feb. 2017.

Stapleton, Andrew. "Stomach Acid Helps Power Tiny, Ingestible Sensors." Cosmos Magazine. N.p., 07 Feb. 2017. Web. 12 Feb. 2017.

Looking to the Light

by Dominique Curtis

I know we're all tired of having our phones die when we have no charger on us or nowhere to recharge them. Well, researchers at McGill University and Hydro Quebec's research institute are on their way to solving all of our problems. The scientists are working on creating a self-charging lithium-ion battery.

Currently our laptops, cellphones, and tablets are powered by lithium-ion batteries. As we all know these device are constantly dying and have to be recharged. This horrible inconvenience is due to the limited energy density of these batteries. This led to the development of portable solar chargers. The issue with these are the difficulty of making them travel-friendly because of how complicated their circuitry and packaging process is.

So researchers and scientists at McGill University and the Hydro-Quebec's research institute are working on developing a hybrid battery. Through their research and studies they discovered that "a standard cathode from a lithium-ion battery can be "sensitized" to light by incorporating photo-harvesting dye molecules" (Paolella, 2017). This means that they can create a charging process that uses light as its energy source.

They are currently working on creating the storage component that will complete the circuit and device. Once they have found a way to do this the device will be complete. Thanks to a grant of $564,000 from the Natural Sciences and Engineering Research Council of Canada it may not take long. The researchers look forward to the ways in which light-chargable batteries will transform portable devices in the future.

Andrea Paolella, Cyril Faure, Giovanni Bertoni, Sergio Marras, Abdelbast Guerfi, Ali Darwiche, Pierre Hovington, Basile Commarieu, Zhuoran Wang, Mirko Prato, Massimo Colombo, Simone Monaco, Wen Zhu, Zimin Feng, Ashok Vijh, Chandramohan George, George P. Demopoulos, Michel Armand, Karim Zaghib. Light-assisted delithiation of lithium iron phosphate nanocrystals towards photo-rechargeable lithium ion batteries. Nature Communications, 2017; 8: 14643 DOI: 10.1038/ncomms14643

McGill University. "Bright future for self-charging batteries: New tech could one day make battery chargers obsolete." ScienceDaily. ScienceDaily, 24 April 2017. <www.sciencedaily.com/releases/2017/04/170424110838.htm>.

Renewable Energy's Implications for the Deep Sea

by Ethan Fukuto

Scientists at the National Oceanography Centre (NOC) in Britain discovered a rich site of rare materials deep underwater. The seamount, located in the Canary Islands, lies about 1000 meters below the surface and is around 3000 meters tall. Samples taken from the seamount contain the substance tellurium at concentrations 50,000 times greater than deposits on land. Tellurium is used in solar panels, and one of the researchers estimated the seamount contains 2670 tons and could provide for 65% of the UK's energy needs. The NOC uses robot submarines to explore and mine the deep sea, and the researchers are now dealing with difficult questions regarding the ethics of deep-sea mining. Deep-sea mining is not a common practice, and its broader implications are thus under-debated and relatively under-studied. MarineE-tech, a research project of the NOC, is currently studying seafloor mineral deposits, the largest and least-explored source of materials, and is working to reduce the impact of extracting minerals. Mining on seabeds would cover smaller areas than on land and would yield richer ores, however marine life is at immediate risk. Plumes of sediment, for instance, could greatly affect the surrounding environment and kill vast swathes of marine life. During the expedition to the seamount, the NOC conducted an experiment which pumped out sediment-filled water while robotic sensors took measurements downstream. Early results suggest that suspended sediment was difficult to detect one kilometer away from the plume, assuaging some fears about mining's impact.

A previous study by the NOC concluded that marine life could recover within a year of mining, but few species would reach their previous population levels. Researchers also note high diversity of single-celled organisms known as xenophyophores in metal-rich nodules, which form shell-like structures that provide habitat for other marine life. While mining may not cause extinction, habitats will be destroyed and will lead to little-studied effects. So, while deep-sea mining could prove to be a rich source of resources for renewable energy, much more research is required to fully know the effects.

Sources

MarineE-tech. "MarineE-tech". National Oceanography Centre.
        http://projects.noc.ac.uk/marine-e-tech/
Shukman, David. "Renewables' deep-sea mining conundrum". BBC. 11 April 2017.
        http://www.bbc.com/news/science-environment-39347620

Report Unveils that U.S. Solar Industry Employs More People than Fossil Fuel Industry

by Genna Gores

The U.S. Department of Energy's 2017 U.S. Energy and Employment Report, reveals that the renewable energy industry employs more people than the entire fossil fuel industry (including petroleum oil, natural gas, and coal). The report goes on to compare employment opportunities between 2015 and 2016 for all types of energy within the Electric Power Generation sector, which includes: solar, wind, geothermal, bioenergy, hydropower, nuclear, fossil fuels, and other generation/fuels. It is evident with this report that solar and other renewable energies are a rapidly growing industry with increasing employment opportunities for Americans.

Within the 2015–2016 year the solar industry has increased by 25 per cent and added 73,000 jobs, which puts solar alone at 374,000 jobs, making up 43% of the Electric Power Generation sector. Wind energy also increased 32%. Greatly exceeding the 187,117 jobs in the American fossil fuel sector which makes up only 22% of the work force, and only an 11% increase in jobs between 2015 and 2016. While the fossil fuel industry is still adding new jobs, it is not nearly at the same rate as renewable energies.

This report comes at an interesting time because President Donald Trump made it very clear that he does not support the green renewable energy industry, and in his energy manifesto said that he will utilize"'America's $50 trillion in untapped shale, oil, and natural gas reserves, plus hundreds of years in clean coal reserves.'" Mr. Trump seems set on boosting fossil fuels despite the fact that industries, such as coal, have seen a net generation decrease of coal sources by 53% over the past 10 years, and solar had an increase of over 5000 per cent. [[http://www.independent.co.uk/news/world/americas/us-solar-power-employs-more-people-more-oil-coal-gas-combined-donald-trump-green-energy-fossil-fuels-a7541971.html]

The facts are obvious that the renewable industry is one worth investing in, but with the Trump presidency many people are worried that market will collapse. There is frustration in the community because reports, like the one from the Department of Energy, demonstrate that renewable energy is one that employs people in general, but more specifically, Americans. The future for green energy is up in the air right now, and hopefully renewable industries will survive the next four years because their job growth is critical.

[https://www.energy.gov/sites/prod/files/2017/01/f34/2017%20US%20Energy%20and%20Jobs%20Report_0.pdf]

http://www.independent.co.uk/news/world/americas/us-solar-power-employs-more-people-more-oil-coal-gas-combined-donald-trump-green-energy-fossil-fuels-a7541971.html

Drones are the Newest Technology in the Solar Market

by Genna Gores

In Cook County, Oregon the largest solar farm in the U.S., Gala Solar Power Plant, will open at the end of 2017. The solar developer, SunPower, developed the solar farm through special drones that map topography, take aerial photos of the landscape, and use custom software technology to pick the best locations for the panels. By using drones to survey new land, companies can spend 90% less time in this stage of solar farm development. These drones work by sending data to SunPower's software, which then compares options for farm layout. Then the company can find the optimum panel layout to maximize the amount of energy on the least amount of land. Beyond saving time, these drones scan for hundreds of layout factors typically overlooked by human engineers. With this technology setting up a solar farm will be quicker and more precise.

On top of better methods for planning, solar companies can also save money on labor due to a new fleet of cleaning robots. These robots will clean the panels at night in 10 hours—a job that usually takes 100 hours for three human workers. This type of innovation, plus the SunPower drones, helps lower the overall cost of solar farms. This will make solar more feasible for the energy market by creating a more competitive energy source. At this time it is important for solar companies to find ways to be more marketable because the current political climate around renewables is detrimental. Solar will lose many subsidies it has had in the past, so with the use of drones it is possible that solar panels will be cheap enough to overcome this loss of funding.

[https://www.fastcoexist.com/3068562/drones-are-making-solar-farms-way-more-efficient]

[https://www.theguardian.com/sustainable-business/2017/feb/26/drones-robots-solar-power-plant-energy]

Sub-Saharan Solar Startups Lighting the Way

by Siena Hacker

According to a December 2016 CNN article, about 68% of people in sub-Saharan Africa live without electricity. Over 90% of Africans, however, have access to cellphones. A new startup, Off-Grid Electric, allows villagers in Tanzania and Rwanda to pay for low-cost solar with their cellphones. The startup, supported by Elon Musk's SolarCity and private equity investment firm, Helios, targets communities that are not connected to an electrical grid. These communities largely rely on burning kerosene, which has negative health effects (such as neurological or kidney damage) and can only emit weak light. Off-Grid Electric already powers about 125,000 homes utilizing rooftop solar panels and a lithium-ion battery. Homeowners pay 10% of their solar package upfront and then utilize their cellphones to pay an additional $7 a month after. According to a December 2016 Forbes article, the $7 a month base package includes enough energy to power four lights and charge a cellphone. This is a reasonable fee compared to the $4 a month most villagers pay for their supply of kerosene or the $400 required to simply connect to an electrical grid. A $20 a month plan is available for those homes that desire more energy, while small business can choose a larger, commercial package to suit their needs as well. Off-Grid Electric is not simply driven by profit; the company aims to truly help the people they are serving. Therefore, if a customer successfully pays the monthly $7 fee for three years, Off-Grid Electric will gift them the solar panels package and no longer charge a monthly fee.

Another startup, ARED, is also attempting to utilize solar innovation to do good works in Africa. According to a January 2017 Disrupt Africa article, ARED provides recharging station kiosks powered by solar energy with a lithium battery. Its mission is to reduce unemployment for all Africans; thus, the franchise business model focuses its employment initiatives on women or the disabled. The transportable kiosks allow customers to pay to recharge their cellphones or other small electric devices. ARED has recently started to integrate purchasable Wi-Fi into their kiosks. There are currently 25 kiosks in Rwanda and the company has plans to bring the franchise to Uganda in March. Both of these startups are helping to power big changes in sub-Saharan Africa in the hopes of ensuring a bight future!

CNN.com x http://www.cnn.com/2016/12/15/africa/off-the-grid-tanzania-rwanda/
Forbes.com http://www.forbes.com/sites/nishthachugh/2016/12/30/what-can-you-do-with-7-a-month-in-africa/#6b0d301b3306
Off-Grid Electric http://offgrid-electric.com/innovation/#finance
Disrupt-Africa.com http://disrupt-africa.com/2017/01/solar-startup-ared-empowers-rwandans-with-business-in-a-box/

Subcutaneous Solar Cells with the Potential to Power Pacemakers

by Siena Hacker

Every seven to eight years, people with pacemakers must endure surgery to replace the battery. Scientists in Switzerland are aiming to obviate the risk, cost, and inconvenience of this surgery with new mini-solar panels that can be placed under human skin. According to a January 2017 Wall Street Journal article, the subcutaneous solar cells are still in development but were very successful when placed under the skin of pigs and connected to a pacemaker. The Swiss scientists recently completed a six-month study measuring how much power their solar panels could generate during daily human activity. As the device is not ready to be inserted under human skin, the study was a feasibility test to see if the amount of energy generated during an average day would be enough to power a pacemaker. Thirty-two Swiss volunteers wore an armband around their biceps containing a box with batteries, electronics, and solar panels which were covered with a filter that imitates the light-transmitting properties of skin. Each participant wore the armband for one week during each season of the year and kept a log documenting weather conditions and daily activities. Participants were instructed to wear the armband over their clothing and to only cover the box when they covered their neck. The Swiss scientists believe that the neck is the probable location for a solar implant, which would be connected to the pacemaker by a small wire. A January 2017 Quartz article reports that the average amount of power generated was always much greater than the 10 microwatts required by a pacemaker. This is such a small amount of energy that the scientists believe the necessary energy could come from artificial light source, such as a common flashlight. The Swiss scientists would also include a tiny battery to store energy for dark periods or to power an alarm when stored energy dips too low. Scientists do not expect that a properly functioning battery will ever go below the energy threshold, and were encouraged that study participants in the age 65 or older group actually generated the most energy. This was likely because these participants were retired and had more time to spend outside. Thus the group most likely to need a pacemaker already demonstrates the best potential to have a successfully functioning solar panel. Though this study only tested the theoretical ability of subcutaneous solar cells, the rate of technological advancement in the solar field seems to suggest that solar powered pacemakers might soon become reality.

@WSJ @qz #Solar panels under human skin? Swiss researchers may soon have a new way to power #pacemakers.

Wall Street Journal: https://www.wsj.com/articles/a-solar-panel-implanted-inside-your-body-1483654391

Quartz: https://qz.com/877351/study-tiny-solar-panels-under-your-skin-are-ready-to-power-the-next-generation-of-pacemakers-and-medical-devices/

Sunny Times for Nevada's Renewable Energy Ahead

by Siena Hacker

Nevada assemblyman Chris Brooks, a democrat on the 2017 Session Natural Resources, Agriculture, and Mining Committee, recently proposed a new bill that would mandate higher levels of renewable energy usage in the Sate. The bill, AB 206, would require that the state's renewable portfolio standard (RPS) reach 80% by 2040. According to Brooks, the bill will aid in diversifying the state's economy and creating high-quality jobs with good wages. The state has a preexisting goal of reaching a 25% RPS by 2025. According a February 2017 Futurism article, the bill would start incrementally increasing Nevada's RPS goal in 2018 by 4% each year until 2030. Starting in 2030, utility companies would have to rely on renewable sources for 50% of their energy production. Brook's bill, which has received great support, is one of many recently introduced measures that promotes renewable energy growth in the state. Many new renewable energy measures, according to a February 2017 Las Vegas Review-Journal, focus on expanding the state's solar capacity since it has a very high solar potential. Development of the solar industry in recent years has created many jobs for Nevadans. However, the approval of Senate Bill 374 in 2015 detrimentally impacted the growth of the solar industry in the state. The bill imposed less generous net metering rates which decreased the financial incentive for investment, resulting in the loss of more than 2,500 rooftop solar installation jobs in 2016 alone. However, the state is now trying to get back on track with its previous solar industry growth. According to a Las Vegas Sun article published in February 2017, a new 50 megawatt solar plant reached commercial operation status in February. Furthermore, Apple and the NV Energy company recently reached an agreement to build 200 megawatts of additional solar by 2019. According to a January 2017 PR Newswire article, NV Energy already produces 491 megawatts of solar energy which supplies customer throughout the state. In total, NV Energy provides energy services to 1.3 million Nevadan customers and operates more than 1,900 megawatts of renewable resources. However, it will take much more imitative on the part of NV Energy and other utility companies to help the state reach AB 206's proposed goal of a RPS of 80% by 2040.

Las Vegas Sun: https://lasvegassun.com/news/2017/feb/14/nv-energy-flips-switch-50-megawatt-solar-plant/

Futurism: https://futurism.com/nevada-pushes-for-80-renewable-energy-by-2040/

Las Vegas Review-Journal: http://www.reviewjournal.com/news/politics-and-government/nevada/nevada-rallies-push-increased-renewable-energy

PR Newswire: http://www.prnewswire.com/news-releases/nv-energy-announces-solar-agreement-with-apple-300396620.html

Project Solar Sheds Light on Potential
by Siena Hacker

Many people associate Texas with big oil companies and gasoline dependence, but a recent report from Google's Project Sunroof sheds light on Texas' potential for solar energy. According to the report, three of the top ten cities with the most solar potential— Houston, San Antonio, and Dallas—are located in Texas. Houston ranks first among the top ten cities, and has an estimated 18,940 gigawatt-hours of potential annual rooftop solar generation. However, potential for rooftop solar is by no means limited to Texas, or even just the stereotypically sunny southern states. According to the report, states like Pennsylvania, Maine, and Minnesota have an estimated 60% suitability for solar rooftop photovoltaics. The project has concluded that, given our current technology, 79% of all the buildings surveyed are technically viable for generating solar energy. Google's Project Sunroof was founded in August of 2015 and has already surveyed the solar potential of over 60 million US buildings in 50 states. The free software gives users who enter their address estimates on what rooftop solar might cost them and what their long term savings would be.

However, there are limitations to Project Sunroof's methodology. The software uses Google Earth's aerial imagery to gauge solar potential, but the data may be imprecise or outdated. Project Sunroof also fails to take into account the various economic factors that might affect a homeowner's decision to utilize solar photovoltaics. For example, net metering rates, tax credits, available financing, and state renewable energy targets can all influence consumer decisions. Though Project Sunroof includes available economic data in its solar price quote it does not currently account for all factors. Some solar providers have partnered with Project Sunroof to provide the next step for consumers. If users choose the "See Solar Providers" button, the website will direct them to various solar companies in their area. These companies can provide the missing details that Project Sunroof lacks and can get a more holistic idea of solar savings and costs.

Alba Energy: http://albaenergy.com/2017/03/texas-has-great-solar-potential-according-to-google

Electrek: https://electrek.co/2017/03/14/googles-project-sunroof-data-79-of-us-rooftops-analyzed-are-viable-for-solar-is-yours/

Greentech Media https://www.greentechmedia.com/articles/read/theres-vast-untapped-potential-for-solar-rooftops-in-the-us-says-google /

Prime Time for Amazon to Order Solar

by Siena Hacker

The online shopping magnate Amazon has greatly benefited from technological advances in robotics, drones, and the internet. The company is now taking on solar as its next venture. According to a Seattle Times article, Amazon will install solar panels on 50 of its facilities by 2020. Amazon estimates that up to 80 percent of a single location's power could be supplied by onsite solar energy generation. The company has 124 domestic warehouses, distribution centers, and regional sortation centers and 146 facilities abroad. By the end of 2017, Amazon will have converted its 15 biggest U.S. shipping and sorting centers to solar. According to CleanTechnica, these 15 centers will have an output capacity of 41 megawatts. A single megawatt of solar electricity can power 164 homes on average, so by the end of the year Amazon is saving power that could benefit 7,000 homes. The benefits from Amazon's solar panels should pay for the cost of installation over the course of their lifetime. The company also generates positive PR from the press the installations generate. Greenpeace recently reported that Amazon had fallen behind Facebook, Google, and Apple in creating renewable energy projects. Some believe that Amazon's foray into solar is an attempt to compete with its rivals for a better public perception. However, the switch to solar also makes economic sense for the company. CleanTechnica reports that rooftop solar power is cheaper than retail electricity in most locations. The company also benefits from the decreasing cost of solar installation and materials. Kara Hurst, Amazon's director of sustainability, recently said that the "cost of technology and the increase in availability has been a contributor" to this undertaking. The company is also expanding its Career Choice Program, which provides financial assistance to employees pursuing new skills, to include training in photovoltaic installation. Amazon's actions demonstrate their outstanding commitment to solar for the foreseeable future.

Clean Technica: https://cleantechnica.com/2017/03/06/amazon-jumps-solar-big-league/
Seattle Times: http://www.seattletimes.com/business/amazon/amazoncom-plans-big-
    solar-power-rollout-at-warehouses/

Solar Shutters Opening New Possibilities
by Siena Hacker

Despite its name, Swedish Solar is actually based in Orlando, Florida. The company was founded by a Swedish expat after he went through process of purchasing solar for his own home. He found that the cost of installation would be higher because his Spanish-style house had a multi-faceted tile roof instead of the more common gable roof. Potential solar customers often face obstacles that render the installation process impossible because of physical limitations or cost constraints. Roofs might face the wrong way, have a wrong slant, or have an insufficient load-carrying capacity. Homeowners might also worry about structural damage to their roof during and after the installation process. Installing traditional rooftop solar panels can void the warranty of new homes, while other homeowners might simply find the appearance of rooftop of solar unattractive. Additionally, some consumers might already have maximized the amount of solar PV on their roofs but still wish to have more generation. For consumers who are unable to install solar on their roof, or have already maximized their rooftop installation, Swedish Solar offers Bahama inspired shutters using monocrystalline solar technology. The company sells custom-fabricated aluminum alloy and stainless steel frames that hold anywhere from one to four solar panels. The shutters, which can be installed on the outside of any building, come in two sizes depending on window size and placement location. For homeowners worried about the aesthetic of the shutters, Swedish Solar states that all wires and controls are hidden in the installation process. The motorized shutters can be controlled remotely using digital devices, such as smartphones, to adjust to the most efficient angle for solar generation. In addition to generating renewable energy, Swedish Solar's window awnings shade homes from excess light and heat. The shutters can also be fully closed for privacy or protection during stormy weather.

Swedish Solar: http://www.swedishsolar.com/the-company/
Clean Technica: https://cleantechnica.com/2017/04/15/swedish-solar-florida-company-
    brings-solar-replace-shutters/

World's Largest Solar Farm: Why Now?
by Parker Head

Tom Phillips' article on China's construction of the world's largest solar panel farm is a useful primer on the various perspectives surrounding the project. China's grandiose scale of renewable energy initiatives can be read as a strategic move to increase their global soft-power. This is prescient in a world where many global superpowers regard climate change as a serious threat. Phillips acknowledges this upscaling as stratagem when he notes its concurrence with the election of a U.S. President who is a climate-denier. But a single solar farm, no matter the grandeur of its sobriquet, cannot nullify the environmental impact of an entire nation, and issues of curtailment – energy produced that does not reach the grid—undercut the potential reformative power of China's green energy production. According to a *New York Times* article also on the new solar farm, 19% of China's wind energy produced in the first six months of 2016 was curtailed, compared to "negligible" amounts of energy lost in the U.S. And while China's current enthusiasm for energy reform is a hopeful sign, long-term commitment is necessary for real change. If current initiatives are only political maneuvers made in a global climate that is changing with the succession of a Trump presidency, then the difference Phillips reports between "*a* climate leader but not *the* climate leader" will be felt in their, the initiative's, long-term ineffectiveness. Is a China-as-world-climate-leader that unimaginable in a world of said radical change?

The citizens of China should not be left out of a discussion of the country's turnaround. With increases of toxic-smog and mass evacuations displacing people as "smog refugees", public discontent over pollution has risen with China's emissions.

Considering domestic crises such as these, China's ideological shift is not only politically expedient but humanitarianly necessary. With the capacity to generate 850MW, enough power to fulfill the energy needs of ~200,000 homes, China's new solar farm stands to make considerable improvements for its citizenry.

[www.theguardian.com/world/2016/dec/21/smog-refugees-flee-chinese-cities-as-airpocalypse-blights-half-a-billion].

[www.nytimes.com/2017/01/05/world/asia/china-renewable-energy-investment.html].

Phillips, Tom. 2017. China Builds World's Biggest Solar Farm in Journey to Become Green Superpower. The Guardian. Jan 19, 2017.
www.theguardian.com/environment/2017/jan/19/china-builds-worlds-biggest-solar-farm-in-journey-to-become-green-superpower#img-1

Phillips, Tom. 2017. Smog Refugees Flee Chinese Cities as 'Airpocalypse' Blights Half a Billion. The Guardian. Dec 21, 2016.
www.theguardian.com/world/2016/dec/21/smog-refugees-flee-chinese-cities-as-airpocalypse-blights-half-a-billion

Michael Forsythe. 2017. China Aims to Spend at Least $360 Billion on Renewable Energy by 2020. Jan 5, 2017. www.nytimes.com/2017/01/05/world/asia/china-renewable-energy-investment.html

Tom Phillips Twitter:
https://twitter.com/tomphillipsin?ref_src=twsrc%5Egoogle%7Ctwcamp%5Eserp%7Ctwgr%5Eauthor

Michael Forsythe's Twitter:
https://twitter.com/PekingMike?ref_src=twsrc%5Egoogle%7Ctwcamp%5Eserp%7Ctwgr%5Eauthor

Morocco Harnesses the Energy of the Sun
by Dena Kleemeier

Morocco is now a global leader in sustainable energy development; however, currently Morocco's foreign energy exceeds 95%. This is possible because the largest solar thermal panel plant in the world, using parabolic reflections to heat molten salt to generate steam, called Noor (meaning 'light' in Arabic) and is being built in Ouarzazate, Morocco. The country predicts that by 2030, over half of it's energy will be renewable, 14% of which will be solar. Furthermore, the project is projected to offset 240,000 t/year of CO2 emissions.

The project is being completed in three phases, Noor I,II, and III. Noor I was started in 2012 and will be 160 MW, upon completion it will have cost roughly 500m € (540m USD) and is being financed by the African Development Bank. The Noor II project will start in 2017 and will be 200MW, will cost 200bn €, and will be financed using 80% debt and 20% equity. Noor III will start in 2018, and will be a 150MN project. Noor I uses a wet cooling system, as part of the steam turbine cycle. The water required for the plants is sourced from a nearby dam (Mansour Eddabhi). The technology licensor for Noor I, II, and III is Sener, and Noor I is equipped with SENERtrough cylindrical parabolic troughs, and will have a molten salt capacity of three hours (II and III will have a capacity of seven hours).

The project is being built by a consortium including, TSK Electrónica y Electicidad, Acciona Infrastructuras, Acciona Ingeniería, and Sener Ingeniería y Sistemas, Sener and Sepco III, 5 Capitals Environmental and Management Consulting (5 Capitals). Although it is advertised that Noor will generate 1,060 jobs per year, in reality, engineers and other workers are being hired by the contractors, thus citizens in Ourazazate aren't being hired to work on the project. Many citizens report that they know very little about the project. Though many Morroccans are qualified for these engineering positions, they are still given to foreigners. This is an economic flaw in the plan and shows that the land is being taken advantage for it's exposure to the sun, as opposed to benefitting locals.

"Noor Ouarzazate Solar Complex." Power Technology. Kable, 2017. Web. 04 Feb. 2017.
Douglas, Danielle. "Morocco harnesses the power of the sun." | Al Jazeera. Al Jazeera, 03
    Feb. 2017. Web. 04 Feb. 2017.

Beautiful, Affordable, Integrated, Tesla's Solar Roofs
by Dena Kleemeier

Beautiful. Affordable. Integrated. This is the argument that Elon Musk makes for the new solar roofs being released by Tesla. Tesla joined with Solar City to produce solar roofs that are aimed to revolutionize solar panels. Musk placed particular emphasis on the aesthetic aspect of the panels as the panels produced are designed to be completely hidden and integrated into the roof itself, invisible from the street. Additionally, Musk states that consumers should have access to install the solar roofs at a cost lower than a normal roof and cost of electricity combined, as most solar panels end up being.

The solar roof is meant to fill battery packs, called the Powerwall 2. These were also newly released alongside the solar roofs and have the capability to power lights, sockets, and refrigerator of a four-bedroom house for a full day. They cost $5,500 and have a 50 Kw power output. The Powerwall 2 stores energy allowing consumers to use energy/electricity at any time with or without sun, and in times of a power outage.

The roof tiles themselves are made with 6 inch textured glass templates that are three layered. The first layer consists of a highly efficient solar cell, followed by a colored louver film that blends the cells into the roof while providing them with exposure to the sun above, with tempered glass covering the three pieces. The tiles are created using a hydrographic process 'using techniques from the automotive glass business' (Musk). In his showcase, Musk played a video in which large weights were dropped on various roof materials including the Tesla tempered glass. All but the Tesla glass shattered upon impact of the weights illustrating the durability of the material.

The solar panels are an exciting innovation for the aesthetics of solar panels and the solar panel industry; however, there are still many questions. With the individual tile panel system, each panel seems complicated to individually install, though Tesla has was able to install the panels in the in experimental functioning homes. Once the panels are public we will see how affordable they are, and how long they take to pay off for the average consumer.

"Tesla Solar." Tesla. N.p., n.d. Web. 07 Mar. 2017.

Pollinator Habitats in Solar Fields

by Genevieve Kules

A few weeks ago I had the opportunity to work with a group of middle and high school students on creating a pollinator habitat to attract bees and butterflies and other pollinators to a farm in Visalia, CA. I was thinking about energy and how pollination has to have an effect on energy, and sure enough I found my connection in news about my home state of Maryland. Pollinator habitats are being planted in solar fields for numerous reasons, among them are to revive the dying population of pollinators, to improve crop yields, to improve scenic views that have been disturbed by solar fields, and to make the soil richer and more stable to protect from runoff into the Chesapeake bay.

While these reasons all sound great, one journalist warns against limited policy standards for the native seeds being planted. She writes that the seeds meant to attract pollinators could be the ones killing them if standards aren't set in the solar industry, meaning that large corporations cannot be trusted without standards. Some seeds marked as poisonous to bees could be overlooked and planted there. She encourages solar companies to follow in the footsteps of the Maryland General Assembly to require approved and non-toxic plants for the pollinator habitats.

One thing I did not know before writing this post is that Maryland has lost over 50 percent of their wild and managed pollinators, and pollinators make possible a third of the food we consume. Combining pollinator habitats with solar farms appears to be a good way forward for both the production of energy—especially for the chicken industry—and for agriculture, as long as we don't poison the pollinators in the process.

Levitsky, Steve. "Solar Offers More than Clean, Low Cost Energy." Delmarva Daily Times. USA Today, 15 Apr. 2017. Web. 25 Apr. 2017. <http://www.delmarvanow.com/story/opinion/columnists/2017/04/15/solar-offers-clean-low-cost-energy/100463142/>.

Raindrop, Bonnie. "Solar Sites Could Accidentally Create Bee Killing Fields." Delmarva Daily Times. USA Today, 23 Apr. 2017. Web. 25 Apr. 2017. <http://www.delmarvanow.com/story/opinion/columnists/2017/04/23/solar-pollinator-friendly-habitat/100751654/>.

Apple Partners with NV Energy on Solar Energy Project

by Nina Lee

On January 25, 2017, NV Energy and Apple announced that they plan to partner up and build 200 megawatts of solar energy by 2019. The agreement is estimated to provide hundreds of jobs to Nevada residents, as well as help Apple reach its renewable energy goals. Apple joined the global renewable energy initiative RE100 in the fall of 2016, even though 93% of its global operations were powered by renewable energy as of 2015. Their goal is to become 100% powered by clean energy sources.

While plans are not underway just yet, NV Energy plans to file an application with the Public Utilities Commission of Nevada to sign a power purchase agreement for the solar power plant, ultimately hoping to hit completion by early 2019.

NV Energy, Inc. comprises of two subsidiaries, Nevada Power Company and Sierra Pacific Power Company. It currently provides over 1,600 megawatts of renewable energy to 1.3 million customers in Nevada, coming from various renewable energy projects including 19 geothermal energy resources, 13 solar energy sources, six hydro-plants, and one wind farm. NV Energy currently provides 491 megawatts of solar power to their customers. After this partnership, that number will increase to over 529 megawatts.

Apple has actually worked with NV Energy before, building a solar panel farm together in 2013. In addition to their partnership, Apple will also contribute up to five megawatts of power to NV Energy's subscription program for residential and commercial customers.

From here, Apple's future plans for solar energy hopefully will include installing solar panels of the rooftops of their Campus 2 headquarters in Cupertino, California. Plans for this projects completion are estimated to be around early 2017. In addition, Apple has already made plans to sell the excess energy generated by these solar panels. They have applied a subsidiary called Apple Energy LLC to the US Federal Energy Regulatory Commission.

https://www.nvenergy.com/company/mediaroom/newsdetail.cfm?n=136958
http://appleinsider.com/articles/17/01/25/apple-nv-energy-forge-deal-to-build-200mw-reno-data-center-solar-farm
http://www.pv-tech.org/news/ferc-documents-show-apple-has-started-a-solar-energy-company
http://www.pv-tech.org/news/apple-commits-to-100-renewable-energy
http://www.pv-tech.org/news/nv-energy-to-build-200mw-solar-with-apple
http://www.prnewswire.com/news-releases/nv-energy-announces-solar-agreement-with-apple-300396620.html

A Bright Future: 10 Year Solar Deployment Plan for Utah

by Nina Lee

In a February 2, 2017 press release, Utah Clean Energy and Salt Lake City, Utah announced their partnership in a 10-year plan aimed to make solar energy more accessible and widespread across the city and state. This plan, dubbed "A Bright Future," is one of the mere fifteen initiatives in the country to receive funding from the U.S. Department of Energy's Solar Market Pathways Initiative. Utah hopes to utilize solar energy, their fastest growing energy resource, to get ahead in the nation's solar energy market.

Utah has experienced tremendous growth in the solar energy sector over the past decade. According to the Wasatch Solar Team, the state only had a recorded 76 rooftop solar installations as of 2006. By 2016, that number rose to several thousand households, producing an estimated total of 140 megawatts of solar energy, 5,894 solar-related jobs, and $300 million in economic benefit to Utah's overall economy. A Bright Future hopes to continue this rising trend and encourage the use of solar energy by removing unnecessary legal restrictions on rooftop solar panels, creating more roof space suitable for installation, and providing more financing options, especially for those on fixed incomes. Salt Lake City and Utah Clean Energy have already collaboratively created the Solar Permitting Toolbox, a service that helps local governments streamline the process of approving solar panel installations.

The most controversial part of this 10 year plan is estimated to be its proposal to evolve the Utility Business Model. The initiative prioritizes the consumer and public utilities, hoping to eventually implement mass adoption of customer-deployed resources (such as solar energy) in order to reduce the amount of money spent on non-solar energy. However, as Utah is an overall conservative state, this proposal has causes business concerns. Both consumers who will continue to purchase energy and the power providers will not be monetarily benefiting at all since the profits gained from this plan will solely go towards the new electricity sales.

http://www.utahbusiness.com/slc-unveils-10-year-solar-plan/
http://www.deseretnews.com/article/865672424/Utah-Clean-Energy-SLC-unveil-10-year-
        solar-deployment-plan-for-Utah.html

Sistine Solar Gives Panels a Make Over
by Kieran McVeigh

All signs point to renewable energy sources like solar to be the energy of the future. Solar panels in particular offer many innovations compared to the traditional power plants of yesteryear; one of these innovations is that theoretically each house could have its own set of solar panels. While there are many logistical hurdles to this, one in particular is that many people do not like how solar panels would look on their roof and therefore choose not install them. Enter Sistine Solar, a new solar panel company that has figured out how to print coating on top of solar panels, that changes the appearance. They currently have many different patterns that can be printed, including patterns designed to match a roof's appearance. The remarkable thing is that with a new technology that the company is keeping under wraps this print does not affect the efficiency of the solar panel. Sistine solar explains that the coating functions in part by being able to trick the human eye with the minimal amount of light reflected, making the panels appear the right color or design that is printed on them.

Up until now Sistine Solar has primarily focused on partnerships with larger companies, as they can print many different things on a solar panel including company logos, or even in some tests, an image of one of the founder's face. Recently these solar panels have become available to the public and have begun to be installed on residential houses, primarily in Massachusetts and California. Sistine Solar claims that they have over 200 residential home owners who are interested installing their solar panels on their homes.

Looking to the future, Sistine Solar suggests they are interested in more company partnerships and in using their solar panels as energy generating billboards. The founders suggest that their panels could be put on top of things like bus terminals, displaying advertisements, while producing energy. This seems like a particularly good idea to me as it makes use of existing structures, and would not change the appearance at all. It seems Sistine Solar offers huge potential of generating a lot of power in a discreet way.

Matheson, Bob. "Solar panels get a face-lift with custom designs." MIT Energy Initiative. MIT,
        23 Feb. 2017. Web. 06 Mar. 2017. <http://energy.mit.edu/news/solar-panels-
        get-face-lift-custom-designs/>.
Barry, Sean. "MIT startup Sistine Solar debuts camouflaging photovoltaics." Construction
        Dive. N.p., 01 Mar. 2017. Web. 06 Mar. 2017.
        <http://www.constructiondive.com/news/mit-startup-sistine-solar-debuts-
        camouflaging-photovoltaics/437164/>

Solar Panels May Help Save Extinct Species
by Nadja Redmond

In the Kutch region in India, scientists plan to use solar energy to revitalize long extinct coral species native to the area. One popular technique utilized around the globe to assist with regenerating coral and creating underwater gardens involves using solar energy to create an underwater structure from "biorock" on which new coral can grow. This base is made from limestone deposited onto a steel skeleton.

A steel structure of any shape or configuration is put onto the sea floor. Cables connect this structure to the solar panels, which float on the surface of the water. Low voltages of electricity—less than 12 volts—are channeled to the steel structure, which prompt a chemical reaction in the seawater similar to electrolysis. Calcium carbonate (limestone) minerals then deposit onto the structure creating the base. Once this is complete, divers and scientists will provide the coral, collected in fragments from other areas where the desired species are not extinct. This coral will attach to the new structure, and begin to grow while also developing the symbiotic relationship with zooxanthellae, a tiny organism which lives in coral and gives it its color—the coral provides shelter and photosynthesis compounds, while the algae produce oxygen and clear waste. The electrical currents will continue throughout the coral's development process; coupled with the limestone base, the coral could grow nearly 20 times faster than normal and have a better chance at survival. In this way, both species will benefit from sunlight.

The Gulf of Kutch contains one of the four major coral reefs in India, but currently only 30% of the coral in Kutch thrives. Coral reefs have very diverse ecosystems, providing food and shelter to millions of species; revitalizing the Kutch reef will be good for the water ecosystem overall. Reefs around the world are dying or reduced to nothing because of pollution, ocean acidification, and human activities. A renewable energy source to help maintain and rebuild reefs will hopefully reverse some of these effects.

http://www.biorock.org/content/method
http://www.biorock.org/content/method
http://www.hindustantimes.com/india-news/scientists-to-use-solar-energy-to-regenerate-locally-extinct-corals/story-oOUY1JVySiJoJlwgRoSFFL.html

New York Thrives while Sunshine State is Behind on Solar
by Yerika Reyes

Although Florida is known to be the Sunshine state, it does not use as much of its generous sunshine for solar energy as might be expected. The Solar Energy Industries Association ranks the state third in the United States for rooftop solar potential however only 12th in terms of installed capacity. There has been some growth in the commercial solar sector, nonetheless, the number of Florida households getting solar panels is not projected to be over 100 a year until 2021, a measly rate for a state with more than 20 million residents. Compared to states in the Northeast states and out west in California, Florida's renewable energy is very clearly behind. Additionally, Donald Trump's administration has been slaughtering any federal policy related to climate change, and the president is expected on Tuesday March 28, to sign an executive order undoing Obama's clean power plan. These federal actions mean the tasks of emissions cuts now depend more heavily on cities and states. Currently, Florida derives less than 1% of its electricity generation from solar.

In many states, a solar company can lend panels to a homeowner and then sell the cheap power generated directly to the owner. Florida does not allow this. Additionally, homeowners are not able to sell their generated solar power to anyone else, such as a neighbor or tenants. Florida requires anyone providing a utility must be able to provide power 24 hours a day. Unfortunately, only the state's enormous monopoly utilities, such as Florida Power & Light, can meet this requirement, therefore households are barred from this sort of third-party ownership.

Jim Kallinger, chairman of the Florida Faith & Freedom Coalition believes solar will continue to be a niche industry as long as Florida maintains its cheap electricity rates. "It's hard for solar PV [photovoltaic panels] to compete with other energy sources," he said to a reporter in the Guardian. "I think solar is more of a novelty for folks who might have some disposable income. Yes, we have a lot of sunshine in Florida but we have a lot of cloud cover. In other states, you have the government intervening more, but here we have pushback against that sort of thing."

Northeastern states may not be immersed in sunlight as much as Florida but have high ambitions for solar. This push for solar in New York is guided by retail politics of electricity prices and a growing acceptance of climate change. New York's governor, Andrew Cuomo, has said warming temperatures are costing the state "not only in dollars but already in lives", while his Florida counterpart, Rick Scott, will be forever burdened by reports that he banned public servants from uttering the words "climate change."

Substantial city and state tax breaks, combined with the dropping cost of solar, are starting to reap dividends in New York City. Installed solar capacity has swelled 800% in the past five years, with state efforts to get panels on to churches and schools. In August, Cuomo announced that utilities would be required to source half of the state's electricity from solar and wind by 2030. Florida does not have a comparable goal. It is obvious that the state support is needed for Florida to begin to flourish in the way New York has.

http://www.seia.org/sites/default/files/Florida.pdf
https://www.theguardian.com/environment/2017/mar/26/trump-executive-order-clean-power-plan-coal-plants
http://www.cnbc.com/2017/02/22/new-york-sees-almost-800-percent-growth-in-solar-power-over-five-years.html
https://www.theguardian.com/us-news/2017/mar/27/solar-power-florida-new-york-renewable-energy-policies

Capturing the Solar Spectrum

by Mary-Catherine Riley

In the last day of 2016, *MIT Technology Review* listed an innovation in solar thermovoltaics as one of the five "Biggest Clean Energy Advances in 2016." This technology eliminates the theoretical maximum of 32% efficiency that is present when converting solar energy into electricity using a single layer silicon cell. This advancement captures previously unused solar energy, such as invisible ultraviolet light and infrared wavelengths, to utilize the full solar spectrum and potentially double the theoretical limit of efficiency.

A MIT lab created this new solar cell that converts sunlight into heat before transforming it into energy using solar thermovoltaics (STV). This technology traps heat from the sun before it can reach the solar cell. As the heat is captured and re-emitted as light, carbon nanophotonic crystals convert the heat to color wavelengths that are beamed at maximum efficiency for absorbance by the solar cell

The benefit of this technology is that the photonic devices produce emissions based on heat and not light and are therefore unaffected by brief changes in the environment such as clouds blocking the sun. Furthermore, when this technology is coupled with the correct thermal storage mechanism, this system could provide continuous power. Nevertheless, this technology is at an early-stage of demonstration and will not be available until researchers increase the laboratory-scale experimental unit to production size and scientists are able to build the cell more economically.

MIT has been a hotbed of solar technology development, including their improvement of the Perovskite solar cells. The continuous advancements in solar technology in the Perovskite and solar thermovoltaics cells as well as other solar technologies will make solar a no-brainer for energy investors and will continue to prop solar as the energy choice of the future.

(http://news.mit.edu/2016/hot-new-solar-cell-0523).

(https://cleantechnica.com/2016/05/31/new-solar-cell-breaks-ceiling-theory-goes-window/).

(https://www.technologyreview.com/s/603275/the-biggest-clean-energy-advances-in-2016/).

Casey, Tina. "New Solar Cell Breaks Efficiency Ceiling, Theory Goes Out Window." CleanTechnica. Important Media, 31 May 2016. Web. 12 Feb. 2017.

Chandler, David L. "Hot New Solar Cell." MIT News. MIT News Office, 23 May 2016. Web. 12 Feb. 2017.

Temple, James. "The Biggest Clean Energy Advances in 2016." MIT Technology Review. MIT Technology Review 2016, 29 Dec. 2016. Web. 12 Feb. 2017.

Solar Energy to Propel Rockets

by Mary-Catherine Riley

In April of 2016, NASA announced a $67 million contract with Aerojet Rocketdyne to advance a Solar Electric Propulsion (SEP) system for future space travel. SEP engines convert solar energy into electricity. This electricity is than used to accelerate ionized propellant at extremely high speeds. The project hopes to double thrust capability compared to current electric propulsion systems and increase fuel efficiency by 10 times the current chemical propulsion capacity. The 10% increased efficiency allows shuttles to power robotic and crewed missions well beyond low-Earth orbit and enable further exploration, ferry cargo, and resupply already underway missions. Aerojet Rocketdyne's current contract is to test the largest and most advanced SEP system on their Asteroid Redirect Mission which is designated to capture an asteroid and place it in orbit around the moon by the mid 2020s.

Electric propulsion has been researched since the 1950s and was utilized in missions such as the 2011 Dawn Spacecraft which orbited and explored the giant protoplanet Vesta and is now exploring the dwarf planet Ceres. A large challenge in SEP technology is that as the shuttle moves deeper into the solar system and further from the sun, it becomes more challenging to efficiently capture the sun's light. Therefore, SEP research is combined with the advancement of solar array tech so the engine can work in all solar environments. The advancement in SEP technology includes advanced solar arrays, high voltage power management and distribution, power processing units, and high-powered Hall thrusters.

In this innovation, NASA is utilizing more green sources of energy because it is more efficient and allows for further space exploration than using other nonrenewable sources. However, I am in no way championing NASA as a beacon of environmental sustainability as its rocket launches create plumes of greenhouse gases and once the rockets fall back to Earth, they deteriorate into the sandy soils and chemically contaminate soil and ground water. Nevertheless, this SEP advancement is refreshing as it is a more environmentally conscious technology that furthers space exploration.

(http://usatoday30.usatoday.com/tech/science/space/2011-07-31-nasa-environmental-cleanup_n.htm).
(https://www.nasa.gov/mission_pages/tdm/sep/index.html)
http://dawn.jpl.nasa.gov).
Calandrelli, Emily. "NASA Invests $67 Million into Solar Electric Propulsion for Deep Space exploration." TechCrunch. TechCrunch Network, 22 Apr. 2016. Web. 26 Feb. 2017.
"DAWN." Jet Propulsion Laboratory California Institute of Technology. NASA, n.d. Web. 26 Feb. 2017.
Mohon, Lee. "Solar Electric Propulsion (SEP)." NASA TV. NASA, 13 July 2015. Web. 26 Feb. 2017.
Waymer, Jim. "Space Program's Environmental Cleanup Could Take Decades." USA Today. Gannett Satellite Information Network, 31 July 2011. Web. 27 Feb. 2017.

Brazil's Emergent Solar Power Energy Market
by Sara R. Roschdi

Brazil's primary source of renewable energy comes from biofuels and hydroelectric power yet since 2009 their solar energy market has increased, making them the leading solar energy market in Latin America and one of the fastest growing markets in the world. Brazil sought out alternatives to hydroelectric dam energy following a drought in 2001 that drastically reduced the energy produced and due to growing concern over the social and environmental effects of building the dams. The Brazilian Federal Energy Regulatory Agency (ANEEL), introduced new initiatives making solar energy more accessible, by allowing people to set up co-operative grid-connected solar energy in which everyone in the community is financially rewarded for the energy produced by the grid. The non-profit, Revolusolar is using this co-operative model to bring green energy to people who live in under resourced favelas in Brazil. Presently several non-profit organizations are pushing for the government to give tax breaks for loans and equipment needed to produce solar energy. They are calling to receive tax breaks comparable to the 15% discount that commercial retail companies, who are the primary consumer of the country's energy resources, receive on their electricity expenses. The Brazilian Solar Power Association (Absor), reports that solar panels have the potential to produce over double the residential demands, and thus decrease the country's dependence on hydroelectirc power. *Telesur* reports that in January 2017 solar power became cheaper than the cost of fossil fuels in Brazil. The increase in solar energy production is aiding the country to sustain energy production during periods of decreased rainfall, allowing them to retain energy prices competitive with with global energy markets. Brazil is being hit with the effects of climate change through prolonged periods of droughts and shifts in their ecosystems and are combating this by becoming one of the globe's fastest growing solar energy producers.

Figueiredo, Rafael, and Larry Pascal. "New Developments in Brazil's Solar Power Sector." Renewable Energy World. N.p., 18 Feb. 2016. http://www.renewableenergyworld.com/articles/2016/02/new-developments-in-brazil-s-solar-power-sector.html

Araújo, Heriberto. "From the Favelas: The Rise of Rooftop Solar Projects in Brazil." Technology and Innovation. Guardian News and Media, 24 May 2016. Web.https://www.theguardian.com/sustainable-business/2016/may/24/favelas-solar-energy-projects-brazil

Johnston, Adam. "13 GW Of Solar Projects In Upcoming Brazil Auction, 22 GW Of Wind." CleanTechnica. N.p., 14 Aug. 2016. Web.https://cleantechnica.com/2016/08/14/13-gw-solar-projects-upcoming-brazil-auction-22-gw-wind/

Alexander, By Rebecca. "Renewable Energy Is Now Cheaper Than Fossil Fuels." News | TeleSUR English. N.p., 10 Jan. 2017. Web.http://www.telesurtv.net/english/news/Renewable-Energy-Is-Now-Cheaper-Than-Fossil-Fuels--20170110-0018.html

Soliculture: Solar Panels Powering Greenhouses

by Chloe *Soltis*

Soliculture is an American start-up that develops and produces electric solar panels for greenhouses. Greenhouses are an efficient method of food production because they use less water and pesticides and have a longer growing season; however, they can be significantly more expensive to operate due to the electricity that is needed to light, heat, and cool the space (UC News). Fortunately, Soliculture has developed solar panels that can be placed on the roofs of greenhouses that generate electricity and support plant growth.

Soliculture's solar panels are transparent, magenta in color, and use LUMO technology to produce energy. The top of the panel is covered with silicon photovoltaic strips and a layer of luminescent material is applied on the bottom side of the panel (Soliculture). When sunlight hits the panel, light-altering dye in the luminescent layer changes the light's color from green to red. This is important because red light is best absorbed by plants for photosynthesis (UC News). Next, some of converted red light travels through the panel to the crops in the greenhouse to support growth and some is consumed by the silicon photovoltaic strips, which increases the solar cell's power and therefore the amount of energy that the panel produces (Soliculture).

Glenn Alers, Soliculture's CEO, is excited for the future of solar panels being used to grow food. He explains that farmers are attracted to using solar panels due to federal tax breaks and the decreasing costs of the solar panels themselves (UC News). Currently, greenhouses are commonly used to grow tomatoes and cucumbers, but with decreasing solar panel costs, he believes that many different crops such as lettuce and strawberries will begin to be grown in greenhouses as well. Solar-powered greenhouses are also expected to be popular in countries such as Canada that have short growing seasons and expensive electricity rates (UC News).

Murdock, Andy. This solar greenhouse could change the way we eat. UC News. March 21, 2017: https://www.universityofcalifornia.edu/news/solar-greenhouse-could-change-way-we-eat.

Soliculture. http://www.soliculture.com.

Azuri Technologies: Using Solar Panels to Bring Electricity to Rural Africa

by Chloe Soltis

600 million individuals in Sub-Saharan Africa do not have access to the electric grid and instead use candles, kerosene, and disposable batteries as their primary sources of energy (Rural Reporters and Azuri). Unfortunately, many of these individuals primarily burn kerosene for energy, which can have negative health and environmental consequences. However, Azuri Technologies, a company based in Cambridge, England, has brought clean, affordable energy to Sub-Saharan Africa by pairing mobile technology with solar panel energy systems.

Azuri's PayGo energy system allows for individuals to install solar panels on the roofs of their homes for a small fee. These solar panels produce up to 8 hours of energy a day that customers can use for lighting, mobile phone charging, and powering small appliances such as TVs. The customer uses a mobile phone to pre-buy power and can pay-as-they-go until the Azuri system is paid off, which typically takes

18 months (Azuri). Then, the customer can decide whether she would like to upgrade her unit if she has more power needs or keep her current system forever. One of the most important components of the Azuri solar system is that its electricity is priced to cost less than what the customer was previously paying for kerosene so she will begin to save immediately after the system's installation.

Azuri has found that their customers have greatly benefitted from their solar panel systems. 97% of customers reported that their children studied more and only 17% still used kerosene after the system's installation (Azuri). Azuri's solar panels have especially helped entrepreneurs, who can now keep their businesses open later into the evening and sell more products. It has also allowed for families who farm to spend more time tending to crops and livestock during the day and then use the evening hours for chores and housework. Azuri Technologies hopes to expand its business in the upcoming years and bring electricity to even more people in Sub-Saharan Africa (Rural Reporters).

Azuri Technologies. http://www.azuri-technologies.com.

Sotunde, Busayo. Azuri: Building a Sustainable Pay-As-You-Go Solar System in Rural Africa. Rural Reporters. March 28, 2017: http://ruralreporters.com/azuri-building-a-sustainable-pay-as-you-go-solar-system-in-rural-africa/.

ColdHubs: Solar-Powered Refrigeration Units in Nigeria

by Chloe Soltis

One of the largest issues in developing countries around the globe is keeping food cold. It is estimated that nearly 45% of food spoils in developing countries before it can be eaten because there was no cold storage available to preserve it (ColdHubs). This matter is especially relevant in Nigeria, where two-thirds of the labor force works in agriculture (CNN). Nnaemeka Ikegwuonu, a Nigerian radio presenter, started ColdHubs to develop a solution to this problem when he recognized that solar power could be used to power refrigeration units in towns that are off the grid.

In 2014, after a few years of research and development, ColdHubs installed 10 of their solar-powered refrigeration units called "Hubs" in ten of Nigeria's busiest markets (CNN). A Hub consists of a walk-in cold-room that is installed beneath a roof of solar panels. The room is built with 4.7-inch-thick insulated panels and is kept at 41 degrees Fahrenheit. In addition, energy captured by the solar panels is stored in high capacity batteries, which are connected to an inverter that powers the refrigerated room (ColdHubs). ColdHubs charges farmers a daily fee of 50 cents for each crate of produce they store in a Hub.

So far, Hubs have been very successful in Nigerian markets. The Hubs allow farmers to keep products fresh for up to 21 days, a substantial improvement over the average 2-day life span of produce (ColdHubs). Ikegwuonu believes that the increased food production will help Nigeria combat other issues such as malnutrition. Eventually,

ColdHubs believes its Hubs can also be used to improve rural healthcare by keeping vaccines and medications cool (CNN). In 2016, ColdHubs planned to install 50 new Hubs with hopes to install many more in the upcoming years.

ColdHubs. http://www.coldhubs.com.

Monks, Kiernon. A radio show host may have fixed Nigeria's worst problem. CNN. April 1, 2017: http://www.cnn.com/2015/12/22/africa/cold-hubs/.

Wind Power

Denmark Runs Entirely on Wind Energy for a Day

by Lauren Bollinger

On February 22nd 2017, Denmark generated enough wind energy to power the entire country for the day. Thanks to particularly windy weather, the Scandinavian nation produced 97 gigawatt-hours (GWh), 70 GWh of which came from onshore wind turbines and the remaining 27 GWh from offshore projects. In total, that amount is enough to power 10 million average EU households for a day.

This feat comes after years of investment in wind energy by the country. Notably, a newly constructed offshore turbine broke a world record in January for the most energy generated by a single turbine in a 24-hour period. Wind power alone accounted for 45% of Denmark's electricity in 2016.

Many European nations have been making similar strides; most notably, Scotland has also been heavily investing in wind energy, with wind turbines contributing to two-thirds of the country's energy needs in February. Additionally, in 2016, the entire United Kingdom was powered without coal for 12 and a half hours, Portugal went four straight days, and Germany went a couple of days on renewables.

Denmark's achievement marks a promising step as Europe as a whole moves to invest in renewable energy. Across the continent, approximately $34 billion was invested in wind energy, and nearly 90% of all new energy construction was spent on renewables. These moves come as the EU Renewable Energy Directive deadline looms near, which requires all member nations to meet a target of 20% energy consumption from renewables by 2020.

Denmark runs entirely on wind energy for a day
http://www.independent.co.uk/news/world/europe/denmark-ran-entirely-on-wind-energy-for-a-day-a7607991.html
Denmark Just Ran Their Entire Country on 100% Wind Energy
https://futurism.com/denmark-just-ran-their-entire-country-on-100-wind-energy/

French company Creates a 'Wind Tree' that Silently Generates Electricity

by Alejandra Chávez

A French company named NewWind has created a 'Wind Tree' that uses micro wind turbines to generate electricity at speeds as low as 4.5 miles per hour. The tree is 26-feet-tall and is fitted with 63 leaf-like blades or 'aeroleaves.' Because these aeroleaves generate a current by rotating a blade moving across a power circuit, micro electricity and multiple turbines are able to work together to capture the lowest winds currents within a 360 degrees radius. Jéròme Michaud-Larivière, the CEO of NewWind, says that the tree is profitable with winds at an average of 7.8 miles per hour annually. In a year, the Wind Tree can generate anywhere from 3,500 Kilowatt hours (kWh) to 13,500 kWh, which is enough to power 15 street lamps of 50 Watts, lighting for 71 exterior parking spaces, filtering a pool of 1,766 cubic feet, or one electric car for 10,168 miles.

The first Wind Tree prototype was developed in 2013 and debuted in Paris in 2015. Currently, the Wind Tree is available for commercial sale at $20,000 to $50,000, depending on the model. It has the potential to be profitable after time since it can power 83 percent of the electrical consumption of a French family household in a year. Traditional models of wind turbines have been deemed loud, large, and too dangerous to be placed in urban areas. The Wind Tree, on the other hand, is silent and could be placed in commercial or even residential areas where regular trees would be. NewWind is even developing other aeroleaf colors, different tree barks, branches, flowers, and bushes to expand options for customers. In the end, the company hopes to be able to connect each tree to generate electricity to be used nearby to make urban areas more sustainable.

[http://www.newwind.fr/en/innovations/].

Barber, Megan. "Urban 'Wind Trees' generate electricity from breezes." Curbed. Vox Media, 14 Mar. 2017. Web. 15 Mar. 2017.

Curbed (http://www.curbed.com/2017/3/14/14914302/wind-tree-turbine-for-sale)

NewWind. Innovations, 2015, http://www.newwind.fr/en/innovations/. Accessed 15 Mar. 2017.

Advancements in Airborne Wind Energy Technology

by Sagarika Gami

The German energy company E. ON announced this week that they will be testing new waters with airborne wind energy technology (AWET). This company is one of the major public utilities in Europe and has recently committed to investing with Ampyx Power in the construction of an AWET site in County Mayo, Ireland, which would test, verify, and demonstrate the operation of its 2 MW Airborne Wind Energy System. Developing a test site is important because it opens the possibilities for collaboration with the many companies working towards implementing AWET technologies, and can also be used as a platform to work with researchers. E. ON's target is to lower the cost for renewable energy, and to work with legislators to introduce this new technology, eventually making it eligible to participate in tendering processes.

AWET, according to Clean Technica, "tethers a specifically configured drone to an offshore platform that is attached to a line, which turns a winch that drives a generator. The autonomous drone aircraft moves in a regular cross-wind figure-8 pattern at altitudes from 200 meters up to 450 meters." The wind energy is harvested by using a fixed wing (similar to kite surfing) that reaches altitudes up to 450 meters. They are cheaper to manufacture and easier to maintain then wind turbines, and thus could transform the global offshore wind generation market. They are also easier to deploy in deeper waters, an important asset for countries such as Portugal, Japan, and the US. The nine countries that share a border with the North Sea, Belgium, Denmark, France, Germany, Ireland, Luxembourg, the Netherlands, Norway, and Sweden, decided last year to invest in improving infrastructure to support offshore wind.

This is the culmination of a 5-year effort. The project is highly important, seen through a 2015 study by the European Wind Energy Agency, which found that offshore wind power is a costly venture and work needs to be done if it can have a viable long-term future.

Winddaily.com
        (http://www.winddaily.com/reports/German_power_company_examining_new_w
        ind_energy_options_999.html)
Cleantechnica.com (https://cleantechnica.com/2017/04/11/e-invests-innovative-drone-
        based-airborne-wind-energy/)
Elp.com (http://www.elp.com/articles/2017/04/e-on-to-demonstrate-airborne-wind-
        technology-in-ireland.html)
Upi.com (http://www.upi.com/German-power-company-examining-new-wind-energy-
        options/9381491906813/)

Hummingbird Turbine a Safer Alternative for Fellow Winged-Ones

by Parker Head

Wind energy harvesting methods have had difficulty branching into the domestic market, especially urban domesticity. The widely used three-rotor model has a large sweep making it impractical for crowded and vertical living. Wind turbines reputation as being harmful to avian life, cited as killing 10,000–40,000 birds annually [http://science.howstuffworks.com/environmental/green-science/wind-turbine-kill-birds.htm], is also a concern for many potential users, with most city dwellers shying away from the prospect of having a bird-blender located on their balcony or roof.

But the Tunisian green energy company Tyer Wind is hoping to create a new market of domestic wind turbines with their hummingbird-inspired Tyer Wind turbine. The Tyer turbine differs from conventional ones in its compact size, each rotor, or more accurately, wing, only 5.25 feet in length, and compact movement. Instead of the broadly sweeping propeller rotation most turbines employ, the Tyer turbine mimics the circular flapping motion of a hummingbird. This motion is supposedly more efficient than the propeller method and, because of its mitigated sweep, safer for birds. The compact area of its rotors' sweep also makes it ideal for higher-density wind farming, allowing the turbines to be placed closer together resulting in a more efficient use of land.

This is the creator's, Anis Aouini, second foray into convenient domestic wind energy capture systems. The hummingbird turbine follows his trajectory of compact and bird-friendly wind energy: his last invention being the Saphonian bladeless turbine, a sail-like device to be affixed to balconies and roofs in dense urban areas where larger prop turbines are not feasible.

Tyer Wind is also looking to scale-up the hummingbird model for largescale off-shore or on-shore usage, the benefits of higher density placement of turbines and fewer bird deaths hopefully exponentially improving at an industrial scale.

[http://newatlas.com/tyer-wind-turbine-hummingbird-wings/47517/].

Anderson, John. 2017. Residential turbine design inspired by hummingbird wings. New Atlas. Jan 25, 2017. http://newatlas.com/tyer-wind-turbine-hummingbird-wings/47517/

Williams, Adams. 2012 Saphonian bladeless turbine boasts impressive efficiency, low cost. New Atlas. November 7, 2012. http://newatlas.com/saphonian-bladeless-wind-turbine/24890/

Layton, Julia. Do Wind Turbines Kill Birds? How Stuff Works. N.d. http://science.howstuffworks.com/environmental/green-science/wind-turbine-kill-birds.htm

Insect-Inspired Wind Turbines

by Parker Head

A team of researchers led by physicist Vincent Cognet at the Paris-Sorbonne University recently found that wind turbines outfitted with rotors inspired by insect wings generated 35% more power than identical turbines outfitted with the conventional rigid rotors.

Increased efficiency in wind energy harvesting is more complex than simply getting turbine rotors to spin more quickly. Wind turbines operating at high speeds are actually less efficient, as well as more vulnerable to malfunctions, than those operating at intermediate rates. When high winds cause turbines to operate at high speeds, the rapidity of the rotors' rotation can effectively create a wall. If a group of turbines is spinning fast enough, wind flow can be blocked from reaching the greatest possible number of turbines. The ideal intermediate rate can be reached by adjusting the pitch of the rotor blades in response to the wind. But, due to the rigidity of conventional wind turbine rotors, the pitch of the blades is not adjustable.

Cognet and his team took inspiration from insect wings, which, due to their flexibility, allow for a continual adjustment of pitch for optimal power and aerodynamics for flight. The physicists built three small scale wind turbines: one with rigid blades, one with mildly flexible blades, and one with very flexible blades. The turbine with mildly flexible rotors, those most like the wings of bees or dragonflies, outperformed conventional rigid rotors by 35% in wind tunnel testing. The flexibility allows the turbines to operate efficiently in a wide range of wind conditions. In high winds the blades naturally compensated with a lower pitch angle, being blown forward into a more closed shape. This shape prevents the wall-effect, which occurs with rigid rotors at high speeds, and the bending slows the rotation, keeping the rate of rotation within optimal energy production range. Cognet and the team are confident that the 35% increase seen in the small-scale models will transfer to full-sized turbines, the dilemma now however is finding the material with which to build the full-size bending blades.

[http://www.sciencemag.org/news/2017/02/wind-turbines-inspired-insect-wings-are-35-more-efficient].

Shultz, David. 2017. Wind turbines inspired by insect wings are 35% more efficient. Science. February 14, 2017. http://www.sciencemag.org/news/2017/02/wind-turbines-inspired-insect-wings-are-35-more-efficient

Cognet, Vincent et al. 2017. Bioinspired turbine blades offer new perspectives for wind energy. Proceedings of the Royal Society A: Mathematical and Physical Sciences. February 15, 2017. http://rspa.royalsocietypublishing.org/content/473/2198/20160726

Science Magazine's Twitter: https://twitter.com/sciencemagazine
The Royal Society Publishing's Twitter: https://twitter.com/RSocPublishing

Artificial Island Devoted to Generating Green Energy

by Parker Head

Even with the advancements in wind energy production and the global rise in wind turbine construction, one drawback has consistently hindered turbine proliferation, people do not necessarily like to be near them [http://www.renewableenergyworld.com/articles/2013/02/wind-farms-a-noisy-neighbor.html] Màr Mack reports on a joint effort being proposed in Europe that would allow for substantial wind energy production while keeping the turbines themselves out of the way.

The Danish company Energinet recently signed an agreement with the Dutch and German company TenneT outlining the preliminary logistics of building an island in the middle of the North Sea devoted to the production of wind energy. The proposed island would be situated in such a position that it could deliver power to the Netherlands, Denmark, Germany, UK, Norway and Belgium. The island is projected to be able to produce between 70,000 to 100,000 MW, servicing about 7,000 wind turbines built on and around it. The construction of an island is not only a way to remove turbines and power stations from people's backyards, but also a way to create a connection between power grids, allowing countries to share electricity.

The island is slated to be functional before 2050 in order help participating European countries meet their Paris agreement pledges. But, beyond that, not much else is known with regards to the island's completion. The projected costs of constructing the island are staggering, the foundations for the island alone costing around 1.45 billion dollars. But, should multiple countries work together to raise the necessary funds and indeed complete the island, it is estimated that around 80 million Europeans would benefit, being serviced with the wind energy produced there.

[https://thenextweb.com/eu/2017/03/13/check-out-europes-crazy-clean-energy-plan/#.tnw_G5WZSVmQ].

Mack, Màr Màsson. 2017. Check out Europe's crazy clean energy plan. TNW. March 13, 2017. https://thenextweb.com/eu/2017/03/13/check-out-europes-crazy-clean-energy-plan/#.tnw_G5WZSVmQ

Casey, Zoë. 2013. Wind Farms: A Noisy Neighbor? Renewable Energy World. February 21, 2013. http://www.renewableenergyworld.com/articles/2013/02/wind-farms-a-noisy-neighbor.html

Màr Màsson Mack's twitter: https://twitter.com/mm_maack

Offshore Wind Farm Industry Takes Off in the United States

by Genevieve Kules

The offshore wind farm industry appears to be growing despite the current political disinclination towards environmentally friendly energy initiatives. In 2016 Deepwater Wind created the US's first offshore wind farm off the coast of Rhode Island's Block Island consisting of five turbines. In January of 2017 Deepwater Wind submitted permits for approval of fifteen turbines off the coast of Long Island, NY. This could only be the start for the construction of over 200 turbines nearby.

Offshore wind farms are far more prominent in Europe, and China has a wind farm with enough turbines to power a small country, but lack of buyers has left many of those turbines unused.

Now, in the United States, offshore wind farms could be a promising energy resource. Many large oil corporations have invested in wind energy and Google says their data centers and offices will be completely run on renewable energy in 2017.

The majority of opponents to offshore wind farms have been citizens interested in protecting their ocean views, but I would imagine the farms will also have an impact on marine life and bird migration. They have the potential to interrupt the fishing and seafood industries as well as high maintenance costs requiring boats and multiple people to fix even the small problems. Though still an important factor to consider, this cost could boost the boating industry and create quite a few jobs. Wind in general is a fairly good source of energy because of its relatively low and consistent monetary cost (at times less than fossil fuels) and its direct production of electricity with no need for processing. Even within this political climate, perhaps wind will take the lead on energy initiatives in the coming years.

Cardwell, Diane. "Off Long Island, Wind Power Tests the Waters." New York Times. N.p., 21 Jan. 2017. Web. 24 Jan. 2017.
<https://www.nytimes.com/2017/01/21/business/energy-environment/offshore-wind-energy-long-island.html>.

"Block Island Wind Farm." Deepwater Wind. N.p., n.d. Web. 24 Jan. 2017.
    <http://dwwind.com/project/block-island-wind-farm/>.
Daniels, Lisa Jan 16 2015 9. "Pros & Cons of Wind Energy." Windustry. N.p., 16 Jan. 2015.
    Web. 24 Jan. 2017. <http://www.windustry.org/pros_cons_wind_energy>.
Watch, National Wind. "Problems with Offshore Wind Farms Not worth It." National Wind
    Watch. N.p., 21 Apr. 2011. Web. 24 Jan. 2017. <https://www.wind-
    watch.org/news/2011/04/21/problems-with-offshore-wind-farms-not-worth-it/>.

BCs Largest Wind Farm, on Saulteau First Nations Land

by Genevieve Kules

The largest wind farm in Canada was completed in late February 2017 by Pattern Development. The farm is located in British Columbia about thirty kilometers north of Tumbler Ridge, on the original lands of the Saulteau First Nation. The wind farm has many positive impacts such as jobs and power. The land on which it is located was previously infected by the pine beetle kill and suffered from deforestation. This means the wind farm has a smaller environmental impact than it would have on untouched land.

Composed of 61 wind turbines, Pattern Development has stated that the project generated 500,000 labor hours to construct and will generate $70 million in its lifetime based on property taxes and other forms of income. It will increase BC's wind power generation by 37% with the ability to power 54,000 homes in the province. It can generate 184.6 megawatts of power, which brings the total wind power generation of BC to 673.6 megawatts. They have a 25 year power purchase agreement with BC Hydro. The project incorporated two different turbines, with different size rotors and different hub heights. Each turbine is situated to maximize wind intake off the ridges of nearby mountains and other environmental factors.

Pattern states that they discussed the project and incorporated input from "First Nations, communities of Tumbler Ridge and Chetwynd, and the provincial government." I began this post by recognizing the Saulteau First Nation and thanking them for the use of their lands. One can only hope that Pattern created this wind farm in a good way, honoring the original protectors of the land and assuring that only they have control over what happens on that land. Previous articles report that some relationships have been formed and treaties signed saying that certain people can pursue environmental construction and initiatives on Saulteau land. But one must be constantly cognisant and appreciative of the land that allows them to live and harness energy to power the surrounding areas.

Lillian, Betsy. "Pattern Brings B.C.'s Biggest Wind Farm To Life." North American
    Windpower. Zackin Publications Inc., 27 Feb. 2017. Web. 28 Feb. 2017.
        <http://nawindpower.com/pattern-brings-b-c-s-biggest-wind-farm-to-life>.
Froese, Michelle. "Pattern Development Completes Largest Wind-power Project in British
    Columbia." Windpower Engineering & Development. WTWH Media LLC, 28 Feb.
    2017. Web. 28 Feb. 2017.
        <http://www.windpowerengineering.com/featured/business-news-

projects/pattern-development-completes-largest-wind-power-project-british-
columbia/?utm_source>.
Mohs, Gordon. "EBA Signs Memorandum of Understanding with Saulteau First Nation."
Tetra Tech EBA. Tetra Tech EBA, July 2013. Web. 28 Feb. 2017.
<http://www.eba.ca/index.php/newsroom/eba-signs-memorandum-of-
understanding-with-saulteau-first-nation-sfn.html>.
"Meikle Wind Is Now the Largest Wind Facility in British Columbia." Clean Technology
Business Review. Progressive Trade Media LTD, 28 Feb. 2017. Web. 28 Feb. 2017.
<http://wind.cleantechnology-business-review.com/news/pattern-completes-
184mw-wind-farm-in-british-columbia-canada-280217-5750499>.

Maglev Wind Turbines

by Byron R. Núñez

Conventional horizontal-axis wind turbines (HAWT) convert only around 1% of wind energy into usable power. This inefficiency can be traced to loss of energy resulting from the friction created as the turbine spins. HAWT-based wind farms are not only expensive to install, operate, and maintain, but they are also noisy and can pose a danger to animals. U.S. based NuEnergy Technologies is developing a new turbine that can eliminate most of the problems found with HAWT. The company is currently producing maglev (magnetic levitation) turbines that are frictionless, wear-free, maintenance free, noiseless, and require no lubricant.

Researchers at the CAS Guangzhou Energy Research Institute have estimated that magnetically levitated turbines can boost wind energy production by 20% when compared to HAWT. NuEnergy Technologies attributes its turbine's efficiency to the moving parts of the maglev turbines that do not touch, creating a nearly frictionless space that allows the capture of energy at speeds as slow as 1.5 meters per second.

The use of maglev technology is not new since it has been used to operate high speed trains in Asia and Europe. NuEnergy Technologies explains, "[t]he Maglev uses a magnetically levitated low revolution per minute (RPM) high-torque power output turbine. The spinning turbine 'floats' on a magnetic cushion, just as the high-speed train 'floats' above the railroad tracks."

NuEnergy Technologies is currently working with other technology partners to offer power plants in sizes that would range from 10 megawatts to 1 gigawatt. The advantage of a maglev wind power plant over a traditional wind farm is significant since it could potentially lower the price of wind energy to less than 5 cents per kilowatt-hour (kWh), on par with coal-generated electricity and only about half the typical cost of wind power. The CAS Guangzhou Energy Research Institute estimates that a 1 gigawatt maglev turbine field would cost $53 million to build and require 100 acres of land, but could supply electricity to 750,000 homes. In contrast, it would cost hundreds of millions of dollars to build a wind turbine field of a similar capacity and require 64,000 acres of and using HAWT.

Popular Mechanics

http://www.popularmechanics.com/technology/infrastructure/g89/8-ways-magnetic-levitation-could-shape-the-future/?slide=6
How Stuff Works Science
http://science.howstuffworks.com/environmental/energy/10-innovations-in-wind-power5.htm
NuEnergy Technologies
http://nuenergytech.com/maglev-wind-turbine/

Flexible Edges for Wind Turbine Blades
by Byron R. Núñez

Wear-and-tear is a considerable problem for any business that seeks to create technologies that are long-lasting, low-cost, and profitable. In the renewal energy sector, constantly having to replace expensive parts can increase the cost of the power generated. For example, in traditional conventional horizontal-axis wind turbines the constant exertion of force damages the blades and increases the cost of the turbines and ultimately the cost of the power they produce. The Risø National Laboratory for Sustainable Energy in Denmark is researching and designing a new flexible edge for the blades that aims to diminish wear-and-tear and other incidental damage experienced by conventional wind turbines.

Risø researchers hope to use a rubber trailing edge that can bend while the blade is rotating, creating a smoother flow of air off the blade and drastically reducing the force placed on the structure. An example of this concept can be found on airplane wing flaps. Researchers explain that, "those flaps alter the wing's shape to offer increased control over lift forces during takeoff and landing. A rubber trailing edge, through similar means, could increase the stability of spinning turbine blades, reducing the amount of stress on the components holding them." However, the big difference is that in aircraft, the movable flaps do not bend to pressure. The flexible trailing edges would provide a continuous surface even when the trailing edges move. The reason for this is that the trialing edge is constructed of elastic material (a pultruded fiber glass reinforced composite) that constitutes an integrated part of the main blade.

By providing the blade with a movable trailing edge, it is possible to control the load on the blade and significantly enhance their lives. The researchers hope to test the rubber trailing edges on a full-scale wind turbine within a few years.

How Stuff Works Science
http://science.howstuffworks.com/environmental/energy/10-innovations-in-wind-power8.htm
Alternative Energy News
http://www.alternative-energy-news.info/elastic-edges-for-wind-turbine-blades/

The Future of Carbon Fiber is Cheaper and More Environmentally Friendly

by Bianca Rodriguez

Carbon fiber is one of many light-weight, strong composite materials used in many areas of manufacturing, including clean energy products. Due to its high strength and low weight, carbon fiber allows wind turbine Engineers to design larger turbine blades. The increased length of the blades allows a single wind turbine to produce a greater amount of power, and increases efficiency, while the lighter weight keeps the blades from failing due to their own weight. Other industries, like aerospace, use carbon fiber to reduce the weight of aircraft, allowing them to use less fuel for any given flight. Carbon fiber can even be used in cars to reduce their weight, therefore potentially increasing their fuel efficiency. However, the production of carbon fiber is very expensive, limiting its use to high end cars like the BMW i8.

The high cost of carbon fiber is attributed to the price of the plastic base material known as a precursor. The most common precursor, polyacrylonitrile (PAN), accounts for 51% of the selling price of carbon fiber. Changing the precursor to a less expensive product will greatly impact the cost of carbon fiber, thus allowing it to be used in a wider range of applications. To lower the cost of the precursor, the US Department of Energy, in partnership with Oak Ridge National laboratories, established the Vehicle Technologies Office (VTO), and the Carbon Fiber Technology Facility (CFTC). One of the many precursors currently in development, Lignocellulosic Sugars, is estimated to cost $1.00/lb. versus PAN which cost $5.00/lb. Many of the precursors currently in development are also derived from biomass as opposed to fossil feedstock. Switching from PAN to Lignocellulosic Sugars can potentially reduce the environmental impact of the manufacturing process. Additionally, if these new precursors can produce carbon fiber that is chemically different from PAN based carbon fiber, it may even allow for carbon fiber to become recyclable.

http://www.nrel.gov/docs/fy16osti/66386.pdf
https://energy.gov/eere/articles/carbon-fiber-and-clean-energy-4-uses-industry
http://www.compositesworld.com/articles/wind-turbine-blades-glass-vs-carbon-fiber
http://www.bmw.com/com/en/insights/corporation/bmwi/concept.html

Harvesting Wind Energy with Invelox Technology

by Chloe Soltis

In September 2011, Dr. Daryoush Allaei founded Sheerwind, an energy start-up focused on using wind power to generate electricity. Dr. Allaei realized that current wind turbines are obsolete in the sense that they must passively wait for wind to operate (Breunig). Dr. Allaei believes that wind's velocity should be accelerated so that electricity can be generated from wind energy in areas that are not suitable for turbines. Therefore, he created the Invelox, a system that can both capture and accelerate wind power.

The Invelox is a large, closed system that is built in a L-shape and contains three main parts: a funnel, a tower, and a Venturi tube (Breunig). The funnel is built on the top of the tower and is used to collect the wind. The funnel can capture wind from any direction. After being captured in the funnel, the wind becomes more concentrated as it passes through the tower's pipes and travels down to the Venturi tube, which contains multiple turbines and rests on the ground. The Venturi tube is smaller than the tower's pipes, therefore the wind's velocity increases when it is squeezed at the tube's entrance. Then, the turbines in the tube can harvest energy as the wind passes through and can ultimately generate electricity.

The Invelox has many advantages over regular wind turbines. The Invelox can produce energy from wind speeds that are as low as 2 meters per second, which means that the Invelox could be used on rooftops in cities in place of power grids. Most importantly, the Invelox is bird-safe because it is a closed system and therefore can be built anywhere. Normal wind turbines kill approximately 573,000 birds every year due to their exposed blades and cannot be used in bird-rich locations as beaches and shorelines (Breunig). Sheerwind currently has five Invelox systems installed all over the world including in Iran, China, and the United States. As of January 2017, the system still needs to be tested by third parties, but Involex has great potential to increase the use of wind power as a clean energy source.

Breunig, Tom. 2016. Wind Harvesting Funnel Takes On Turbines. Cleantech Concepts. Jan 24, 2017: http://www.cleantechconcepts.com/2016/09/wind-harvesting-funnel-takes-on-turbines/.
Sheerwind. http://sheerwind.com.

Icewind: Wind Turbines for Everyday Use
by Chloe Soltis

In 2012, Saethor Asgeirsson founded Icewind, an energy start-up that produces small vertical axis wind turbines to be used for both commercial and residential purposes. Asgeirsson is a mechanical engineer from Iceland, the world's largest green energy producer per capita. Asgeirsson realized the need for smaller and more efficient wind turbines both in Iceland and internationally due to rising energy prices and an increasing demand for clean energy sources. Therefore, he designed and developed small vertical axis wind turbines that could be used for a variety of purposes but also avoided some of the issues that plague conventional wind turbines.

Asgeirsson's turbines are made-up of four curved blades, and when the blades are placed together, the resulting turbine resembles a pinwheel. However, unlike a pinwheel, the turbine is extremely strong because its parts are made up of carbon fiber, stainless steel, and aluminum. It is also a fraction of the size of normal wind turbines and can even be installed on the roofs of homes. The Icewind turbine is designed to generate power in wind conditions as low as 2 m/s and can even slow down when wind conditions are extreme. This is important because conventional wind turbines can catch on fire due to complications with high winds. The turbines also do not negatively affect bird life and can catch wind from any direction. In addition, Icewind turbines are very low maintenance and will last for many years, a great improvement over current turbines in Icleand which have a life span of three years due to the harsh weather conditions. Icewind claims that the turbine's aesthetic design is one of its best features since many individuals have commented on how the turbines look like sculptures and blend in with the natural landscape (Bloomberg).

Icewind is focusing on using the turbines in both commercial and residential settings. The Icewind team envisions the turbines being used in homes on and off the grid and especially in telecom towers and surveillance spots. They are also being tested in cities. Most recently, Icewind installed smaller versions of the turbine on the roof of a bus stop. The turbines powered smart phone recharging stations and advertising boards at the stop. Icewind turbines were made available to the public market in Iceland in 2016 and are expected to be available internationally in 2017.

Icewind. http://icewind.is.

Not Too Windy, Not Too Calm: Capturing Iceland's Winds. Bloomberg. Jan 30, 2017: https://www.bloomberg.com/news/videos/2016-06-20/your-fish-sticks-may-have-been-sliced-by-this-algorithm-ipomonsy.

Ovy, Danny. 2015. Iceland Uses a Unique Technology to Produce Wind Power. Alternative Energies. Jan 30, 2017: https://www.alternative-energies.net/iceland-uses-a-unique-technology-to-produce-wind-power/.

IceSolution: Using Microwaves to Deice Wind Turbines

by Chloe Soltis

IceSolution is a Norway start-up focused on developing technology to deice the blades of wind turbines. It was founded in 2014 to continue the research that its parent company Re-Turn AS started in 2008 as a Eurostar project (IceSolutions). Ice accretion on wind turbine blades is problematic for wind farms in cold regions. In the United States, 65% of all wind turbines are located in areas where icing is possible, and when turbines do ice, they oftentimes have to be shut down and need to be visually inspected before they can operate again (WindPower). Five mm of ice on a blade can decrease the turbine's power output by up to 80% while also potentially causing damage to the blade itself (WindPower). Current technologies indicate when icy conditions are present but do not monitor whether ice has physically formed on the blades.

IceSolution has developed a system that can melt ice off the blades of wind turbines. Their system consists of a transistor microwave unit that moves up and down the turbine's pole and blades that have been covered in a microwave heatable coating called ConCenTrate. The

ConCenTrate coating consists of carbon nanotubes and is painted underneath the white topcoat of the turbines. ConCenTrate absorbs the microwaves emitted from the transistor unit and then releases heat on the blade's exterior surface, which melts the ice (IceSolutions). The system can be installed onto any existing turbine.

While this is an external deice solution, the company is also working on an internal one. They want to build the transistor microwave units into the blades of the turbine so that no equipment will be visible from the outside of the turbine. IceSolutions has also been collaborating with the Norwegian National Center of Expertise for Smart Energy to design antennas that could be used in conjunction with their transistor microwave technology (IceSolutions). In March 2017, their product had not been released to the public market but will hopefully be available soon.

Early, Catherine. Harnessing wind energy in icy climes. WindPower. March 7, 2017:
	http://www.windpowermonthly.com/article/1183991/harnessing-wind-energy-icy-climes.

Ice Solution. http://www.icesolution.no/technology/.

Harper, Nick. Detecting Ice on Wind-turbine Blades. WindPower. March 7, 2017:
	http://www.windpowerengineering.com/maintenance/detecting-ice-on-wind-turbine-blades/.

Renewable Energy from Artificial Trees

by Justin Wenig

Researchers from Iowa State University have developed a wind energy generator that looks like a cottonwood tree. After considering various options to maximize energy production, the researchers built a generator that creates electricity as its artificial leaves sway in the breeze. Though the device produces less electricity than commercial wind turbines, it may be an attractive option for communities concerned with the look of wind turbines. More, by demonstrating that a biomimetic wind energy generator is feasible, a flood of research and investment into exploring the concept of a tree-like generator is probable.

The mechanics of the generator are simple. Researchers placed small, flexible plastic strips inside of the artificial leaves of the generator. As wind moves the leaves, the plastic strips bend and generate an electrical charge. This method of electricity generation, called piezoelectric effects, is not the most efficient method of energy generation, but it allowed the researchers to make a generator more aesthetically pleasing than traditional monolithic wind turbines.

While the generator has an aesthetic advantage over wind turbines, the researchers admit it must improve in efficiency to compete with other renewable energy sources. In Tim Sandle's February 2017 article published in the Digital Journal covering the generator, Sandle explains that the generator does not produce enough electricity to make it a viable solution for most communities. However, Sandle explains, the researchers are optimistic that there may be more efficient methods of transduction to power a tree-like generator that exist.

Given the potential size of the market for a wind energy generator that can blend in with the environment, it seems likely that someone will pick up the researcher's concept and run with it. If so, perhaps artificial trees will become a large source of renewable energy in the future.

http://www.digitaljournal.com/tech-and-science/science/generator-mimics-trees-to-
produce-renewable-energy/article/485069
journals.plos.org/plosone/article?id=10.1371/journal.pone.0170022

World's Largest Windfarm Set for Construction
by Justin Wenig

Developers have begun constructing the world's largest floating windfarm off the coast of Scotland. Statoil, the energy company responsible for the project, hopes to have five six-megawatt wind turbines constructed in the North Sea and producing electricity before the end of 2017. Floating offshore wind energy turbines are an exciting technology with huge potential.   According to The Guardian journalist Damian Carrington, more than 40 commercial floating wind turbine projects around the globe are in the pipeline to be deployed. However, many of the projects have been delayed due to the difficulty of perching a large wind turbine on a floating platform in rugged waters. While building offshore wind farms may not be easy, investors see huge potential in capturing energy from the strong winds above deep water. In fact, the Department of Energy has earmarked 50 million dollars to begin construction on floating wind turbines off the coast of Oregon in early 2017.   According to Carrington, if floating wind turbines prove to stand the test of time and show long-term efficacy, they could be attractive solutions along the coast of California and Portugal. In addition, turbines could also help communities avert the political difficulty of placing loud, monolithic wind turbines near the homes of residents. It is worth noting that there is a dearth of research on the environmental consequences of offshore wind turbines. According to a 2014 study published by researchers at St. Andrews University, marine mammals are attracted to large anthropogenic structures at sea. Naturally, offshore wind turbines may harm marine animals, and have attracted criticism from conservationists and marine biologists alike. However, more research is needed to determine how negative of an impact the farms may have on marine mammals. Regardless, this is an exciting technology with big potential.

http://www.cell.com/current-biology/abstract/S0960-9822%2814%2900749-0
https://www.theguardian.com/environment/2016/may/16/worlds-largest-floating-
windfarm-to-be-built-off-scottish-coast
https://www.energy.gov/articles/3-can-t-miss-technologies-2017-arpa-e-summit
https://arpa-e.energy.gov/?q=slick-sheet-project/robotic-personal-conditioning-device
https://www.washingtonpost.com/news/energy-environment/wp/2016/06/30/this-robot-
is-really-cool-seriously-its-a-rolling-air-
conditioner/?utm_term=.chttps://www.washingtonpost.com/news/energy-
environment/wp/2016/06/30/this-robot-is-really-cool-seriously-its-a-rolling-air-
conditioner/?utm_term=.caed75a869ebaed75a869eb

Typhoon Turbine

by Justin Wenig

Japanese green energy startup Challenergy Inc. has developed an egg-beater shaped wind turbine that can capture energy from a typhoon. Unlike traditional wind turbines, the device can stand straight in extreme wind and rain conditions. Challenergy Inc. engineers estimate that the energy from a single typhoon could power Japan for 50 years, if there were any way to store it. The device could be useful in developed countries with frequent typhoons, such as Japan, China, The United States and The Philippines.

According to an October 2016 article written by Chisaki Watanabe at Bloomberg, the turbine stands 23-feet tall and utilizes three rotating wind-resistant cylinders. As wind whirrs around the cylinders, the cylinders spin and power the generator. The unconventional shape helps the generator capture energy from wind blowing in different directions at variable speeds. While traditional propeller-based wind turbines stop operating when wind speeds exceed 25 meters per second, the egg-beater turbine can withstand wind speeds of up to 80 meters per second. Junko Ogura at CNN has reported that the the device has achieved a 30 percent efficiency level in early tests, a sizeable number considering the amount of energy in a typhoon.

Challenergy has raised almost 4 million yen in crowdfunding and recently received support from the Japanese government. The company has also received several offers from venture capitalists to help speed the time to bring the device to market. With the help of financial backers, Challenergy hopes to have the device up and running by the Olympic Games in Tokyo 2020.

@Bloomberg @CNN great articles on #Challenergy typhoon turbine

WATER/HYDROPOWER

Groundwater Desalination May Be Solution to California's Drought

by Aurora Brachman

Water officials are exploring the desalination of California's brackish groundwater as a potential solution to California's drought. The director of the Oakland-based think tank spearheading the water program believes there is an abundance of available brackish groundwater. The group also says desalination of brackish groundwater is cheaper and has fewer social and environmental impacts than ocean water desalination. This announcement comes a few months before the U.S. Geological Survey is projected to issue an extensive report mapping the aquifers of brackish water throughout the United States.

Groundwater desalination has been successfully used in the Alameda County Water District in the Bay Area, an area which serves 350,000 residents. Their brackish water desalination facility has provided 14,000 acre-feet of water a year since 2003; roughly 40 percent of the water for the district. The newly proposed desalination facilities would be much larger than this, with projections it could supply 23,000 acre-feet of water annually.

Brackish groundwater is 3.5 times less salty than seawater, making it nearly twice as cheap to desalt. The Pacific Institute report estimates brackish groundwater desalination will cost $950 to $1,300 per acre-foot, in comparison to $2,100 to $2,500 per acre-foot to desalinate seawater. The other big appeal to brackish water desalination is it does not endanger marine life and there are few other environmental consequences.

Brackish water has been an invaluable asset to states and even countries that struggle with their water supplies. Texas is one such place; it currently has 46 brackish water desalination facilities. The country of Israel also gets roughly 10% of its water supplies from desalination plants. With these promising examples of brackish water desalination as a viable solution to California's drought, three new desalination plants are under construction and there are plans to construct at least 17 more in the near future.

Bansal, Correspondent Devika G. "Desalination of aquifers offers drought-weary California new hope." The Mercury News. The Mercury News, 06 Feb. 2017. Web. 07 Feb. 2017.

(http://www.mercurynews.com/2017/02/05/desalination-of-salty-aquifers-offers-drought-
        weary-california-new-hope/)
Register, Lauren Williams Orange County. "Desalination in California: Q&A on making fresh
        water." The Mercury News. The Mercury News, 23 Jan. 2017. Web. 07 Feb. 2017.
(http://www.mercurynews.com/2017/01/23/desalination-in-california-qa-on-making-
        fresh-water/)

Glacier Melts and Causes River to Change Course in One Day

by Aurora Brachman

Climate change has caused a glacier to recede far enough that a river in Canada's Yukon territory changed course. The meltwater from the mammoth Kaskawulsh glacier suddenly switched direction. This is believed to be the first time in modern history that this has happened. This phenomenon is known as "river piracy", and typically occurs when a river is diverted towards another body of water by a landslide or dam collapse, never before by a glacier. The water that once flowed into the Sims River and then north to the Bering Sea in the Arctic is now flowing to the Alsek River and into the Pacific Ocean.

This event was published in the Nature Geoscience Journal. The authors have no doubt that this event of "river piracy" is a result of modern climate change. It is possible that this has been a common occurrence during previous periods of climate change in earth's history.

This event of river piracy is believed to have occurred over one day, May 26, 2015. The last time the glacier was visited in 2013, scientists described it's river as "swift, cold, and deep". However, when they returned in 2015, the river the glacier used to drain into was still and shallow. The scientists stated they were surprised to find "there was basically no river".

There is a slim 1 in 200 chance that this event could have been cause by something other than man made climate change. For the scientists who published on this event, there is no question. They stated "Climate change is happening, is affecting us, and it's not just about far-off islands in the South Pacific... climate change may bring new challenges that were not even really thinking about."

Devlin, Hannah. "Receding glacier causes immense Canadian river to vanish in four
        days." The Guardian. Guardian News and Media, 17 Apr. 2017. Web. 18 Apr.
        2017.
(https://www.theguardian.com/science/2017/apr/17/receding-glacier-causes-immense-
        canadian-river-to-vanish-in-four-days-climate-change)
Ellsworth, Barry. "Climate change forces Canadian river to change course." Anadolu Agency.
        N.p., 18 Apr. 2017. Web. 18 Apr. 2017.
(http://aa.com.tr/en/americas/climate-change-forces-canadian-river-to-change-
        course/799295)

Monitoring Groundwater in Africa Brings Greater Prospect to Water Security

by Ethan Fukuto

Hand-pump systems, which extract groundwater via a well, may provide many communities throughout the African continent water security through a mobile sensor technology. Researchers at the University of Oxford have been installing low-cost mobile sensors on hand pumps since 2012. The sensors were initially designed to monitor the motion of the pump to determine whether it was functional or not, allowing for quicker repair times; prior to their installation, detection of failure, replacement, and repair could take over a month. The sensors send data via SMS, providing information on hourly water usage. However, the research team likewise discovered they can use the data from the sensors to determine the supply of water in a given aquifer through analyzing the pump's vibration. Pumps vibrate differently, researchers discovered, depending on the amount of water below. The researchers use machine learning to analyze the high-frequency "noise" of the hand-pump data, mapping out a relationship between pumping action and the depth of groundwater. The amount of groundwater is estimated to be one hundred times greater than that of Africa's surface freshwater supply, and communities and governments are better able to assess supplies and its maintenance through these data. Daily and weekly information, researchers note, is useful for detecting early signs of problems as well as assuring a more secure supply of water within communities. Monthly and seasonal data are useful for the broader management of water. Regulators can more accurately identify sustainable levels of extraction for agricultural, commercial and mining purposes while likewise providing for small-scale, domestic consumption. Three hundred sensors have been installed in Kenya, and have been well-received by communities. The research team has termed this discovery an "accidental" infrastructure, as the sensors exploit the widespread use of communal hand-pumps instead of requiring a major overhaul of communities' infrastructures. If adopted more widely, these sensors can provide security for around 200 million people who depend on groundwater.

Colchester, Farah et al. "Accidental infrastructure for groundwater monitoring in Africa". 2017. http://www.sciencedirect.com/science/article/pii/S1364815216308325
McGrath, Matt. "'Good vibration' hand pumps boost Africa's water security". 24 February 2017. http://www.bbc.com/news/science-environment-39077761

Graphene Membranes Convert Saltwater to Drinking Water

by Ethan Fukuto

Researchers at the University of Manchester's National Graphene Institute have developed a graphene membrane capable of filtering out common salt in water to create potable water. Prior to this development, graphene-oxide membranes have piqued scientists' interests for their gas separating-capabilities, and their ability to filter out nanoparticles, organic molecules and large salts in water. Prior to this new discovery, common salts were not effectively filtered out. To do so, the graphene membrane must have holes less than one nanometer in size. However, graphene membranes swell when placed in water, allowing the smaller, common salts to pass through unfiltered. By placing walls of epoxy resin on either side of the graphene, swelling is blocked. Now, with consistent and small pore sizes in the graphene-membrane, researchers have developed a reliable process for desalinating water. When a salt molecule is dissolved in water, a 'shell' of water molecules forms around it. The salt molecule requires the help of this 'shell' to pass through; by contrast, water molecules can pass through individually. Tiny capillaries are thus able to block the larger salt molecule while allowing the water 'shell' to move through. The research team is working on ways to make this technology as inexpensive as possible to implement it in countries without the financial infrastructure for large plants.

Scalable membranes with consistently-uniform pore size has likewise opened up a number of possibilities in different contexts, with the ability to filter out ions according to size, the researchers note. According to the Institute, graphene can also be used as a coating for food and pharmaceutical packaging to block the transfer of water and oxygen so foods and perishable goods can last longer. These membranes could likewise remove carbon dioxide in power stations, which they note is not in practice at any scale. Along with its desalinating capabilities, graphene membranes may provide safer and more sustainable food and water production and consumption.

Rincon, Paul. "Graphene-based sieve turns seawater into drinking water". 3 April 2017. BBC. http://www.bbc.com/news/science-environment-39482342

University of Manchester. "Graphene sieve turns seawater into drinking water." ScienceDaily. ScienceDaily, 3 April 2017. <www.sciencedaily.com/releases/2017/04/170403193120.htm>.

National Graphene Institute. "Membranes". University of Manchester. http://www.graphene.manchester.ac.uk/explore/the-applications/membranes/

Neonic Pesticides Found in Drinking Water
by Ethan Fukuto

Researchers at the University of Iowa have detected small traces of neonicotinoid chemicals in tap water for the first time. Neonicotinoid chemicals are the most commonly used insecticide in the world; first introduced in the 1960s, the chemicals are applied to the seed coating and are found to be lethal exclusively to insects. The chemical is absorbed by every cell in the plant, making all parts poisonous to insects. The use of these insecticides received public attention back in 2014 for their potential harm to bees, causing the European Union to block their use on flowering crops. In the US Midwest, the insecticide is common in surface water near land used intensively for agriculture. A 2015 study by the US Geologic Survey collected samples from 48 different rivers and streams throughout the United States, and found at least one neonic chemical in 63% of the streams sampled.

During conventional treatment, researchers found that three neonic chemicals—clothianidin, imidacloprid and thiamethoxam—remained in the water, with concentrations from 0.24 to 57.3 ng/L. They found little to no removal of chlothianidin and imidacloprid and only about 50% removal of thiamethoxam. Though these numbers are relatively low, the effect of exposure is currently unknown and understudied. Under the EPA's regulations, neonics are not currently considered a threat to humans. However, a major concern is how these insecticides may be transformed during treatment processes into harmful substances. Clothianidin, for instance, can react with free chlorine and could undergo some form of transformation during chlorination. On the up side, researchers also took water samples from the Iowa City treatment facility, which removed all clothianidin, 94% of imidacloprid, and 85% of thiamethoxam, attributing its success to its granular activated carbon (GAC) filtration system. GAC filtration is a relatively inexpensive process, and researchers were encouraged by the results.

Hladik Michelle L., Kolpin Dana W. (2015) First national-scale reconnaissance of neonicotinoid insecticides in streams across the USA. Environmental Chemistry 13, 12-20.. https://doi.org/10.1071/EN15061

Kathryn L. Klarich, Nicholas C. Pflug, Eden M. DeWald, Michelle L. Hladik, Dana W. Kolpin, David M. Cwiertny, and Gregory H. LeFevre. Occurrence of Neonicotinoid Insecticides in Finished Drinking Water and Fate during Drinking Water Treatment. Environmental Science & Technology Letters Article ASAP. http://pubs.acs.org/doi/abs/10.1021/acs.estlett.7b00081

McGrath, Matt. "First study finds neonic pesticides in US drinking water". 5 April 2017. BBC. http://www.bbc.com/news/science-environment-39504487

McGrath, Matt. "Widespread impacts of neonicotinoids 'impossible to deny". 24 June 2014. BBC. http://www.bbc.com/news/science-environment-27980344

"Tree-on-a-chip" Device: A New Kind of Hydraulic Pump

by Sagarika Gami

Engineers at MIT are working to design a microfluidic device called a "tree-on-a-chip." Trees and other plants pull water up from their roots to their leaves and take the sugars produced by their leaves down to the roots. This process can be seen as a natural hydraulic pump. The exchange of nutrients is processed through a system of tissues, xylem, and phloem. These natural processes have directly influenced the creation of the "tree-on-a-chip" device. The device operates passively, mimicking the pumping mechanism of water and sugars through the chip.

MIT's Associate Department Head for Operations, Anette Hosoi, says that the chip's passive pumping could be used as a hydraulic actuator (a cylinder motor that uses hydraulic power to enable mechanical operations) for small robots. The "tree-on-a-chip" device is useful in this situation because it is both difficult and expensive to make small movable parts and pumps to power movement in small robots. Hosoi explains the goal of the device to be "cheap complexity" within a small-scale hydraulic system. Drawing from the model of xylem and phloem along with the function of a tree's leaves maintains a constant osmotic pressure that circulates water and nutrients continuously.

The chip is made up of two plastic slides with small channels that represent xylem and phloem. The xylem channel is filled with water and the phloem channel with water and sugar. The two slides are separated with a semipermeable material that imitates the membrane between xylem and phloem. Another membrane is placed over the slide with the phloem channel, and a sugar cube is set on top. The chip is hooked up to a tube that feeds water from a tank into the chip. This setup allows for the chip to passively pump water from the tank into the chip and finally into a beaker.

Spacedaily.com
(http://www.spacedaily.com/reports/Tree_on_a_chip_passively_pumps_water_for_days_999.html)

Questions on Beneficial Reuse of Wastewater
by Sagarika Gami

Scientists are questioning "beneficial reuse," the technique of taking wastewater produced from an oil or gas well, treating it, and using it for other purposes. The treated wastewater can be used to water crops, feed livestock, and for other farming processes. This technique is most often used in California as a remedy for the drought, and though it has been used for many years, its use is sufficiently infrequent that there is little known about related health or safety risks. Only 2% of wastewater is treated, while the other 98% is injected into disposal wells underground.

Concerns about water shortages in the future make beneficial reuse appear attractive, but there are still many questions about what is in the wastewater and could it be toxic? Coming from the oil and gas industry, this water holds an array of chemicals, with more than 1,600 chemicals potentially present.

Though some advocates of beneficial reuse claim that we know how to safely treat the water, there is doubt in the scientific community as studies have only been able to detect about a quarter of the 1,600 suspected chemicals. Along with this, there is very little known about how toxic these chemicals may be. Detection methods are difficult to employ, as oil and gas wastewater is extremely salty, sometimes more than ten times saltier than the ocean. These technologies do not work as well in the presence of such high salt content.

Without accurate testing, we are rendered unsure about whether the current treatment processes are effective. The Environmental Defense Fund is leading many efforts to expand this research, but there is a long way to go.

Breakingenergy.com (http://breakingenergy.com/2017/03/24/scientists-question-risks-of-using-oilfield-wastewater-on-food-crops/)

New Solar Power Water Purification System Can Help Millions in Rural India

by Genna Gores

Scientists at University of Edinburgh are developing a solar-powered water-purification system, which could help 77 million people in rural India access safe drinking water. In many rural parts of India there is no governmental system for water treatment, especially for sewage, so many people do not have access to clean water. These scientists are working to develop a way to clean water while keeping the cost and complexity of the system down.

The water filtration system uses a photocatalytic material that, when exposed to light, creates a chemical reaction. This reaction generates high-energy electrons that activate oxygen, which ultimately destroy harmful bacteria in the water. This material is easy to create, so it could be produced locally, and in its simplest form can be placed in a transparent water bottle. Then this water bottle filled with contaminated water, after sun exposure, could kill all harmful microscopic bacteria. After a regular filter could be used to remove any large particles, if need be.

This is an innovative advancement in the fight to give access to clean water in many rural areas around the world. Currently the scientists from Edinburgh are pairing up with researchers at Indian Institute of Science Education and Research in Pune to help roll out a pilot program in a few Indian towns. If this water-purification method works it could mean an off-the-grid system that requires no outside power source. While this will help many people access water there are other ways that water could get recontaminated that need to be addressed. If water is disinfected and then left to sit without a top, comes into contact with dirty hands, or exposed to other sources of contamination it could be dangerous once again. This solar purification could help water access issues, but is only a temporary solution. India, and other countries, faces a larger issue of water infrastructure. Many of these rural areas do not have proper sewage systems or water treatment plants, and it is important that the solar purification system does not minimize the importance of infrastructure advancements.

[http://www.seeker.com/solar-purification-could-provide-clean-water-in-rural-india-2236280454.html]

[http://indianexpress.com/article/technology/science/solar-powered-system-to-provide-safe-drinking-water-in-india-4488067/]

[https://www.fastcoexist.com/3067580/this-ultra-cheap-material-uses-sunlight-to-purify-drinking-water]

California Drought Over

by Cybele Kappos

California's drought had been going on for six years and a state of emergency was announced in early 2014, but recently, Governor Jerry Brown ended the state of emergency for the vast majority of the state. The effects of the drought were devastating. A major reservoir in Northern California called Folsom Lake had been reduced to less than a third of its capacity in 2015. The Governor's announcement was not meant to permanently appease any fear of water supply shortage. It only marked a formal end to a water shortage which has been accompanied by a loosening of many of the state's strict rules regarding water usage due to the drought.

In April 2015, the Governor had ordered a 25% reduction of urban water use. State officials came up with several tactics on how to combat excess water use including setting up rebate programs to allow residents to transition to more water-friendly gardens such as using ground cover materials like gravel.

The state's hydrology improved massively beginning in late 2016 when a series of storms drenched Northern California. The storms continued throughout the winter and major reservoirs swelled to the point that officials had to make releases. The snowpack also made a magnificent recovery. Alongside the end of the drought, the state announced that its snowpack is currently around 160% of what is normal at this time of year.

The only counties that did not have the drought emergency lifted are Fresno, Kings, Tulare, and Tuolumne because the drought affected them most severely. According to a report, Californians managed to slash their water use by more than 22% between June 2015 and January 2017. State officials have released a plan to continue to practice water conservation in the years to come and will retain prohibitions on wasteful water practices. Water conservation must become a way of life for Californians.

Stevens, Matt. California, Drenched by Winter Rain, Is Told 'Drought's Over.' April 7, 2017
https://www.theguardian.com/us-news/2017/apr/07/california-drought-over-jerry-brown-future-climate-change
https://www.nytimes.com/2017/04/07/us/california-drought.html

Low Flow Showerheads in Hong Kong Lowering Water Consumption

by Dena Kleemeier

Having lived in California since the fall of 2015, and experienced the drought and water conservation efforts, reading about the Hong Kong Water Supplies Department (HKWSD) efforts towards carbon conservation and reducing water consumption. I have been consistently encouraged and reminded to be aware of my water usage, particularly with sinks and showers since coming to Southern California as these conservation efforts are well publicized, and I personally have control over my waste. Kwok-wai Mui Ling-tim Wong and Yang Zhou's article in the August 2016 edition of Water, evaluated the impact of low flow showerheads for Hong Kong residents. Thus, the HKWSD created a Water Efficiency Labeling Scheme (WELS) to categorize certain showerheads based on water efficiency, energy us, and CO_2 emissions, then created a water efficiency grades for the showerheads, and then monitored their use. After studying the participant's use of the showerheads, the results indicated that

a showerhead with a lower nominal flow rate can improve water efficiency. Specifically they predict that if WELS-rated showerheads were fully implemented in the city they 'can reduce water consumption 37%, energy use by 25% and CO_2 emissions by 26%. In reading these statistics, I immediately thought of water conservation efforts that exist in California, and other areas that are experiencing a drought. Though I believe that it is an individual's responsibility to conserve water, and not produce excess waste; however if there are systems and technologies in place such as water efficient shower and tap heads that make it very convenient to reduce water consumption, energy use, and CO_2 emissions, California has seen drastic change in their water conservation levels. I believe that this change has been so effective because it removes the human factor, as humans aren't rational and won't always prioritize the environment prior to their own conveniences. As effective changes have been measured in California, a similar trend will likely be seen in Hong Kong.

Wong, Ling-Tim, Kwok-Wai Mui, and Yang Zhou. "Impact Evaluation of Low Flow Showerheads for Hong Kong Residents." Water 8.7 (2016): 305. Web. <http://www.mdpi.com/2073-4441/8/7/305/htm>.

Saudi Arabia Water Desalination Fueled With Thermal Energy

by Dena Kleemeier

The Kingdom of Saudi Arabia heavily relies on the desalination of water to meet their water consumption needs. I experience this first hand when I am home and all the water that comes out of the taps is brackish, and not drinkable. In 2010, 70% of the freshwater withdrawn in Saudi was sourced from non-renewable fossil groundwater, the supply of which has been depleting at a constant rate of 2% per year. In response, Saudi Arabia uses desalination technologies to convert saline water into freshwater, making Saudi the largest producers of desalinated water in the world. Though using desalination has reduced the country's need to draw from natural aquifers, there have been significant financial and environmental costs as a result. Thus, the Kingdom has looked into concentrated solar power (CSP) and thermal energies to fuel their desalination needs,

Saudi Arabia has previously used crude oil to fuel the desalination of water; specifically, in 2011 they used 28 million barrels to desalinate all of the water they distributed. Saudi commonly uses their oil to source their various energy needs, thus making "oil consumption increase by 5.7% per year" (Napoli, 2015). From an economic standpoint, Saudis should prioritize finding a different energy source for desalination as 87% of the Kingdom's revenue is from oil, thus using that source to power desalination is only depleting their resources. Environmentally speaking, the Kingdom's carbon footprint will naturally decrease with finding other energy sources. Thus, the World Bank explored the extent to which CSP can be stored during times of low sunlight to provide 24 hours of power, and concluded that it has high potential to provide energy security and reduce greenhouse gases. By 2015, 57% of output capacity for desalination came from thermal technologies, 40% from reverse osmosis, and the remaining 3% came from advanced desalination. The reason that thermal technology is so effective in the KSA is because it uses a simple distillation process to separate the salt from the freshwater, and other impurities. Secondly it can use waste heat from cogeneration power plants, allowing water and energy needs to be met simultaneously from one utility, and lastly the energy prices are low. With thermal energy as a viable option for desalination, Saudi is taking steps closer to reducing their immense carbon footprint.

"International Journal of Water Resources Development." Taylor and Francis Online. N.p., n.d. Web. 27 Feb. 2017.

Combined Water and Energy plant takes aim at Regions in Drought

by Kieran McVeigh

For several years before the winter of 2016/17 much of California remained in a severe drought. Droughts like this in population centers around the world are a more and more common occurrence with change in climate due to human-related activity. Researchers at MIT set out to design a system that could help alleviate some of the water shortages caused by droughts like these. The team created a system that both creates energy and works to create drinking water from salt water, all while running off of renewable energy.

The system starts with a reservoir located in a coastal mountain range. Water from the ocean is then pumped from the ocean into this reservoir using solar energy or another renewable energy source. Pumping the ocean water uphill serves as a way to store energy. When energy is needed, the water flows down the hill through turbines, which generate energy at times when other renewable energy sources are unable to. This hydroelectric pumped storage plant is combined with a reverse osmosis water desalination plant to provide clean drinking water from ocean water flowing through the turbines. The researchers freely admit that their design is in large part based on a design proposed in 1980s from Kyoto University's Masahiro Murakami, however they say this design was never developed to the extent they developed their design because of how cheap oil was back then. Which just goes to show how much economic pressures shapes innovation around climate change.

Analysis of feasibility of this design showed that it would actually be feasible to build one these plants in Southern California, large enough to serve the energy for 28 million people and that once built this plant would operate at a cost of $5,000 to $10,000 per person served. This finding lends legitimacy to the design, that it is not just a good idea but could be practical for large amounts of desalinated waterin the real world. Moving forward researchers suggests systems like these could help settle the unrest in places where drought causes political turmoil.

Travers, K. (2017, March 28). Combined energy and water system could provide for millions. Retrieved April 04, 2017, from http://energy.mit.edu/news/combined-energy-water-system-provide-millions/

Slocum, A. H., Haji, M. N., Trimble, A. Z., Ferrara, M., & Ghaemsaidi, S. J. (2016). Integrated Pumped Hydro Reverse Osmosis systems. Sustainable Energy Technologies and Assessments, 18, 80-99.

Hydropower in California Surges, but This Means There Will Be Corporate Shifts

by Nadja Redmond

Recent winter storms in California have produced a substantial snowpack across the region and increased lake levels, setting a foundation for major use of hydropower for energy generation this year in the state. This boom, though positive in many aspects, has produced an imbalance institutionally in energy costs, causing a reduction in power prices. This is good news for rate players, but will have a negative outcome for independent power producers (IPPS). Citi analyst Anthony Yuen paraphrased the root of the imbalance in this statement: "The more hydro generation you have, the less [natural] gas generation you should have." He anticipates the year-to-year hydrogeneration on the West Coast to continue to be higher than normal due to the significant snowpack generated throughout the stormy season, with California climbing a little bit closer to matching the hydrogeneration consistently recoded in the Pacific Northwest. At the beginning of March, California's Sierra Nevada mountain snowpack was at 188% of normal statewide snowpack.

A Moody's senior vice president and analyst covering utilities and power explained how competition due to increased use of renewable energy sources, mainly wind and solar in recent years, has resulted in power prices that have dropped. An increase in hydropower, another renewable energy source, will add to this depression. This is a negative for IPP companies, especially those invested in California gas-fired plants.

However, some public power utilities such as Silicon Valley Power, Sacramento Municipal Utility District and others with hydrogeneration assets, who purchased power on the market during the drought years will benefit from this lower cost of power as hydrogeneration increases.

California holds 290 hydro-generation plants, with the largest being the Shasta Dam. At full capacity, the generators can produce 710 megawatts of hydropower, which is enough to power 500,000 homes. At the beginning of March, the dam was at 93% capacity and 130% of the historical average. This trend is repeating in many of the other state reservoirs.

[http://www.cnbc.com/2017/02/23/californias-hydropower-surge-may-be-bad-news-for-these-power-producers.html]

[http://www.cnbc.com/2017/02/23/californias-hydropower-surge-may-be-bad-news-for-these-power-producers.html]

[http://cdec.water.ca.gov/cdec app/snowapp/sweq.action].

Solar and Seawater Farms

by Yerika Reyes

How does a thriving farm in the heart of the arid Australian desert exist with no soil, no pesticides, no fossil fuels, and no groundwater? Sundrop Farms researchers have spent the last six years fine-tuning a system that pipes seawater in from the ocean and desalinates it using a state-of-the-art concentrated solar energy plant. Then the water is used to irrigate 180,000 tomato plants grown in coconut husks instead of soil, kept in a network of greenhouses. Sundrop Farms is a commercial-scale facility located just off the Spencer Gulf in South Australia that began construction in 2014. Today it's producing an estimated 17,000 tons of tomatoes per year to be sold in Australian supermarkets.

There is an increased demand for fresh water around the world and this new way of farming may represent the future of large-scale farming, especially in coastal desert regions that have previously been non-arable. At the center of the farm are the 23,000 mirrors that reflect sunlight towards a 115-meter high receiver tower. The concentrated sunlight produces 39 megawatts, enough to cover the desalination needed by the farm and supply all the electricity needs of the greenhouses. Furthermore, during scorching hot summers, seawater-soaked cardboard lines the greenhouses to help keep the plants at optimal temperature. Additionally, the seawater sterilizes the air, making chemical pesticides unnecessary.

Even with all these innovations there can be some negative aspects. Firstly, there's very little research on whether this sunlight-collecting system has had any damaging impact on other animals in the desert. Similar mirror-based solar facilities in the U.S. are known to incinerate upwards of 6,000 birds per year as they fly in front of the highly concentrated sunbeams to hunt insects.

Additionally, the facility is expensive, it cost around 200 million dollars to begin with, but in the long run the facility should save money compared to the costs of conventional greenhouses that require fossil fuels for power. It is a self-sustaining, cost-efficient design so long as the initial investment can be provided. Facilities similar to the Australian one are already being planned for Portugal and the U.S., as well as another in Australia. Although more research is needed, this could be the farming of the future.

http://modernfarmer.com/2016/10/sundrop-farms/
http://www.sciencealert.com/this-farm-uses-sun-and-seawater-to-grow-vegetables-in-the-desert
http://www.sundropfarms.com/

Indigenous Communities Resist Hydroelectric Dam Projects in Guatemala

by Sara R. Roschdi

The government of Guatemala has approved hydro electric dams to be built on indigenous territories. Fitzpatrick-Behrens reports in the article, "Electrifying Guatemala: Clean Energy and Development" that these hydroelectric dam projects are expected to produce 181 megawatts of energy for the country For indigenous communities like the Ixcán community, these dams mean the pollution of their waters and the corporatization of their sacred lands. Telesur reports that on January 17th, two Indigenous Guatemalan activist, were assassinated by the state as they engaged in a peaceful protest against the building of a hydroelectric dam in San Mateo Ixatan, Guatemala The expansion of these hydroelectric dam projects are a result of foreign investments, that are increasing the funding of energy projects in order to meet the potential energy demands of a growing export economy, since the passing of the Central American Free Trade Agreement (CAFTA). These hydroelectric dams, are being built to reduce the cost of energy, for primarily international businesses and manufacturers. Fitzpatrick-Behren finds in a study conducted by Guatemala's *El Observador* that manufacturers spend approximately 40% of their cost on energy, paying on average 19 cents per kilowatt hour compared to the rate of 10 to 12 cents in countries with competing markets. These foreign investments are leading to the destruction of the environment, increased debt and the displacement of Indigenous and Afro-descended communities. The World Bank is advertising these energy initiatives as promoting environmental sustainability, though using "clean energy" sources while they are simultaneously funding mining projects, the building of highways and the extracting of petroleum across the region. This energy source that is advertised as a green alternative to the use of fossil fuels, is spreading across Latin America and Indigenous communities are resisting to save their territories, their water supplies, and protect Mother Earth.

Fitzpatrick-Behrens, Susan. "Electrifying Guatemala: Clean Energy and Development." NACLA. N.p., n.d. Web https://nacla.org/news/electrifying-guatemala-clean-energy-and-development

Reeves, Benjamin. "Backlash Continues over Hydroelectric Power Projects in Guatemala - ." The Tico Times | Costa Rica News | Travel | Real Estate. N.p., 10 Apr. 2014 [https://nacla.org/news/electrifying-guatemala-clean-energy-and-development]. http://www.ticotimes.net/2014/03/10/social-discontent-swells-over-hydroelectric-power-projects-in-guatemala

Keeanga-Yamahtta Taylor, Amanda Alcantara, Geovanny Vicente, Ivonaldo Leite, and Tortilla Con Sal. "Guatemalan Activist Killed Protesting Hydroelectric Project."

News | TeleSUR English. Telesur, 18 Jan. 2017. Web.
http://www.telesurtv.net/english/news/Guatemalan-Activist-Killed-Protesting-Hydroelectric-Project-20170118-0013.html
[http://www.telesurtv.net/english/news/Guatemalan-Activist-Killed-Protesting-Hydroelectric-Project-20170118-0013.html].

Clearer Waters Ahead for Blue Energy

by Justin Wenig

A captivating article published by an international team of scientists in the August issue of Nature magazine could make blue energy a powerhouse sustainable energy source in the near future.

Blue energy, or osmotic power generation, refers to energy derived from the difference in salt concentration between freshwater and saltwater. At river estuaries, where river water and sea water meet, blue energy can be captured when molecules from the saltwater side move toward the freshwater side and spin a turbine.

Unfortunately, scientists have long struggled to develop a commercially viable generator with a positive return on investment. Case in point, the world's first commercial osmotic power generator, commissioned by a Norwegian company Statkraft, could only produce enough energy to power one-tenth of one electric car battery before it was shunned in 2014. The cost? Ten years and over $100 million lost at sea.

According to Jurica Dujmovic, a columnist at marketwatch.org, the recent findings by EPFL (Ecole polytechnique fédérale de Lausanne) researchers could significantly improve the efficiency of commercial generators, and demonstrated it by building a small prototype generator. By crafting a thin membrane (similar to the intersection between saltwater and freshwater) with a selective nano-pore separating the two liquids, the researchers were able to increase the current flow of water, speeding up the turbine and capturing about a thousand times more electricity per square foot than the failed $100 million Statkraft generator.

What this means is, if scientists can scale the micro-generator pioneered by EPFL researchers, blue energy could vault toward the front of affordable sustainable energy sources. More, it could become an important tool in the growing arsenal of sustainable energy sources and a hedge against vicissitudes in weather that affect solar and wind energy.

It is worth noting that the potential impact of blue energy is contingent on the preservation of estuaries. For estuary conservation activists, the research findings serve as a strong argument against human-made estuary degradation and drainage. To wit, plentiful freshwater basins are needed to produce plentiful blue energy. To some extent, this shifts the conversation on estuary conservation from an ethical one to an economic one.

http://www.nature.com/nature/journal/v536/n7615/pdf/nature18593.pdf
http://www.marketwatch.com/story/this-breakthrough-in-blue-energy-could-change-the-world-2016-08-04

R.E. Pattle (2 October 1954). "Production of electric power by mixing fresh and salt water in the hydroelectric pile". Nature. 174 (4431): 660. doi:10.1038/174660a0.

WaterSeer: The End of Water Scarcity?

by Justin Wenig

A new technology aims to provide a sustainable source of safe water to people living in water-scarce regions. WaterSeer is a device that extracts moisture from the air and condenses it into clean water. Developed by VICI-Labs in collaboration with The National Peace Corps Association and The University of California at Berkeley, WaterSeer is chemical-free, uses no power, and will be priced at $134. The device is scheduled for field trials in late 2017.

According to Karen Graham of the Digital Journal, WaterSeer is planted approximately 8 feet deep into the ground. The underground chamber of the device contains metal sides that are cooled by the soil. Above ground, a lightweight helical turbine spins in the breeze and sends air into a condensation chamber. As the air cools in the chamber, water vapor condenses along the sides, which then drips down into the underground chamber. Clean water is then extracted from the reservoir through a portable pump. Because the above-ground air is always warmer than the metal sides of the underground chamber, the device collects water all the time, even in windless conditions.

The potential impact of this device can not be understated. Water scarcity is a serious issue that affects millions of families around the world. According to water.org, in 2016 one in ten people worldwide lacked access to clean, safe drinking water. In addition, 18,000 people die everyday from a lack of safe drinking water. Further, according to the World Bank, water-scarcity is a major deterrent to economic growth and education in developing countries.

The team behind WaterSeer estimates that a single device has the potential to collect up to 37 liters of water per day. If the device can meet even a fraction of that estimate, it seems that it could solve a myriad of global crises.

http://www.digitaljournal.com/news/environment/waterseer-may-be-the-answer-to-creating-water-self-sufficiency/article/477226
https://www.waterseer.org/
http://water.org/water-crisis/water-sanitation-facts/
http://www.worldbank.org/en/news/press-release/2016/05/03/climate-driven-water-scarcity-could-hit-economic-growth-by-up-to-6-percent-in-some-regions-says-world-bank

Water Filter Made From Wood

by Justin Wenig

In March of 2017, researchers at KTH Royal Institute of Technology announced that they have developed a bacteria-trapping material made from wood fibers. The material can be used to make a portable, low-cost water filter for communities without access to clean water. It also shows promise for use in packaging, bandages, and plasters.

According to sciencedaily.com, the material is a combination of two substances, wood cellulose and a positively-charged polymer. The material traps bacteria by binding the negatively-charged bacteria to the positively-charged material surface. As bacteria are trapped to the substance, the natural antibacterial properties of wood render the bacteria unable to move, and they eventually die. One of the primary advantages of the material is that the bacteria cannot develop resistance to the substance, which is a common problem with many anti-bacterial substances.

The substance can be used to make a water filter for communities without a clean water supply. Since the filter does not have an energy requirement, it could improve access to clean water in communities without a power grid. Further, the filter is eco-friendly and does not leak any toxic material like other cheap, portable on-site water filters do. The team of Swedish researchers hopes to make the filter available to communities in need in the near future. After that, they plan to develop and sell bandages, plasters and packaging, and other substances that need to trap bacteria. With a great degree of versatility and an affordable price tag, the wood-based substance could be used in a variety of ways in the near future.

https://www.sciencedaily.com/releases/2017/03/170320085507.htm
http://www.treehugger.com/clean-technology/wood-fibers-make-cheap-portable-water-
    filtration.html

Biofuels

U.K. Grocery Chain Cuts Back with New Delivery Trucks Powered by Food Waste

by Alejandra Chávez

Waitrose, a chain supermarket in the United Kingdom, has adopted several projects to reduce its food waste. These efforts have proven to be significant because, since 2012, the sixth largest grocery retailer in the United Kingdom has stopped sending food waste to landfills. One way food waste has been reduced is thanks to Waitrose's discounted prices for "imperfect" or bruised produce. The food that is not sold is usually donated to local charities or other "good causes." More recently, in February of 2017, Waitrose has partnered with Compressed Natural Gas (CNG) Fuels to use food waste to also power ten of its new delivery trucks. This is done through anaerobic digestion, which captures biomethane from food waste and funnels that gas into the national gas grid [https://www.fastcoexist.com/3068244/these-grocery-delivery-trucks-are-powered-by-food-waste]. CNG Fuels then takes the gas from the grid and compresses it to track the amount of fuel that the truck uses. This is done so that the fueling stations can "buy" precisely the correct amount of biomethane to fuel the vehicle.

Though there are several benefits to food-based fuel, a *Curbed* article also explored some of its challenges. For one, biodiesel fuel is expensive and the trucks that run on biodiesel fuel are 50% more expensive to buy. Additionally, fully electric trucks that are powered by biodiesel have a difficult time with the weight of the batteries. Unlike a city bus, these large trucks must be on the road for longer hours and the required time to charge — although only five minutes — continues to be a challenge. Regardless, the benefits of food-based fuel are clear: Food-based gas is both affordable and sustainable because it is 35 - 40% cheaper and emits around 70% less carbon dioxide than diesel fuel. Over the course of one truck's lifetime, which is roughly 500 miles of travel, savings are expected to amount to $100,000 [https://www.fastcoexist.com/3068244/these-grocery-delivery-trucks-are-powered-by-food-waste].

Ro, Lauren. "New food waste-fueled delivery trucks will roll out for U.K. supermarket." Curbed. Vox Media, 20 Feb. 2017. Web. 20 Feb. 2017.
Curbed (http://www.curbed.com/2017/2/20/14669484/waitrose-sustainable-food-based-fuel-new-delivery-strucks-cng-fuels)

Peters, Adele. "These Grocery Delivery Trucks Are Powered by Food Waste." Co.Exist. Fast Company, 17 Feb. 2017. Web. 20 Feb. 2017.
Co.Exist (https://www.fastcoexist.com/3068244/these-grocery-delivery-trucks-are-powered-by-food-waste)

Producing Jet Fuel from Kitchen Oil
by Sagarika Gami

Next year, a big state-owned refiner in China will start using waste kitchen oil from restaurants to mass-produce jet fuel. The project will be taken on by a subsidiary of Sinopec: Zhenhai Refining and Chemical, based in Ningbo, Zhejiang province. They will construct a new production plant that can convert 100,000 tons of waste kitchen oil to 30,000 tons of aviation-grade biofuel per year. The fuel produced will be marketed at airlines with international flights, specifically those that fly to countries that charge high emissions taxes.

Biofuel doesn't add extra carbon dioxide to the atmosphere, which is why it is regarded as a source of clean energy. However, critics point out that biofuel degrades in long-term storage. Also, biofuel forms a gel at low temperatures, potentially constricting or blocking fuel flow in pipelines. Also, using kitchen waste products is difficult because they must go through an intensive filtering process to get rid of the vegetable oil, animal oil, various proteins, salt, etc. mixed in it. The filtering process begins by heating the oil to over 350°C to eliminate water molecules. Hydrogen is added to extend the oil's shelf life. The biofuel made by Sinopec has met safety standards.

This project started in 2011, with the first sample synthesized in 2012, and actually used in a commercial flight in March 2015. The firm hopes to begin producing output in 2018. Since 2011, more than 1,500 flights have flown using biofuel, but these were for demonstration. The rise in regulation of greenhouse gas emissions shows the future for Sinopec's biofuel idea, as alternative fuel both makes more financial sense and is kinder to the environment. Air travel is expected to double in the next 20 years, making this project imperative and relevant.

Breakingenergy.com (http://breakingenergy.com/2017/04/03/kitchen-oil-for-jet-fuel-renewable-energy-new-projects-clean-biofuel-jets-international-flights-china-testing-alternative/)

Biofuels Innovations in Hawaii
by Cybele Kappos

A plant operated by Pacific Biodiesel in Keaau, Hawaii is currently producing nearly 13,000 gallons of biodiesel fuel a day. It is the first in the U.S. to be certified as sustainable by the Sustainable Biodiesel Alliance. The fuel is created from waste cooking oils, animal fats, fruit and seeds. Cardwell examines the extent to which supposedly green alternatives to fossil fuels actually reduce carbon emissions.

Starting in 2005, the federal government approved a measure that required biofuels to be blended into gasoline at increasing rates, helping spur its production. Most of that was from food crops like corn and sugar cane which led to controversy. The demand led to farmers clearing land, including rainforests, in order to grow more crops, thereby releasing more carbon dioxide into the atmosphere. The demand for those crops can also drive up the price of food and animal feed. Consequently, producers moved away from the production of ethanol and toward advanced biofuels created from agricultural waste, but these turned out difficult and expensive to produce. The Pacific Biodiesel plant seeks to avoid all the above problems of ethanol and the biofuels. Instead, the plant creates its fuel from local waste products, including restaurant cooking oils and grease and agricultural products such as macadamia nuts that are considered unsuitable for the market. At the plant, the oils are processed and distilled into fuel. Methanol, which is used in the refinement process, is recycled and the company is currently trying to develop markets for the byproducts of the process. The company sells almost all of the resulting fuel to customers in Hawaii. It is striving to reinvent agriculture in Hawaii and is constantly experimenting with new feedstocks. Although it is only a small company, its efforts are worthy of notice.

Cardwell, Diane. Biofuels Plant in Hawaii is First to be Certified as Sustainable. May 13, 2016.
https://www.nytimes.com/2016/05/14/business/energy-environment/biofuels-plant-in-hawaii-is-first-to-be-certified-as-sustainable.html
http://www.utilitydive.com/news/hawaii-biodiesel-plant-is-1st-in-us-certified-as-sustainable/419212/
http://www.businesstimes.com.sg/energy-commodities/biofuels-plant-in-hawaii-first-to-be-certified-as-sustainable

Croton Nut Biofuel's Potential in Kenya
by Nina Lee

Eco Fuels Kenya is a small company based in Nanyuki, Kenya that has begun to explore the option of using croton nuts as a source of biofuel. Croton trees are native not only to Kenya, but are commonplace throughout East and Central Africa. The trees are currently not used for much more than firewood, but their nuts, although inedible, have been discovered to be filled with oil and protein. The harvest period for the croton nut lasts up to six months a year, maximizing production. Eco Fuels Kenya has been researching the potential of a croton nut-based biofuel since 2012 and now hopes to introduce it to the mainstream.

Croton biofuel produces about 78% less carbon dioxide than diesel and can go directly into generators, water pumps, or tractor engines made for diesel. However, it still needs to be processed in order for it to be used in cars. Eco Fuels Kenya has also made an effort to put the byproducts of their fuel production to use. Pressing the croton nuts leaves behind a seedcake paste that can be used to feed poultry, and the husks can be ground up and used as organic fertilizer.

One of the biggest challenges that Eco Fuels Kenya faces is the reluctance of investors and Kenyan government officials to back a biofuel project after the previous failed jatropha biofuel project in 2000. The government had fully endorsed the jatropha project and seized farmers' land in order to create large scale farms in hopes that the resulting biofuel would spur Kenya's economy. The project ended up not being successful, and thus left many farmers with no land and no jobs. With this failed experience in mind, Eco Fuels Kenya actively utilizes a different business model. The company is collaborating with local farmers, currently sourcing from about 5,000 farmers around Mount Kenya and the Rift Valley. All suppliers are located within 100 kilometers of their factory, as Eco Fuels Kenya stands by their promise of benefiting local communities.

http://www.efkgroup.co.ke/
https://www.theguardian.com/sustainable-business/2017/jan/10/croton-oil-biofuels-
kenya-food-security-farming-africa
http://www.cnn.com/2016/12/28/africa/croton-nuts-biofuel-aes/

Pacific Biodiesel Technologies Announces Maui-Based Biofuel Project

by Nina Lee

On February 24, 2017, Pacific Biodiesel Technologies kicked off their latest Maui-based biofuel project by performing a native Hawaiian blessing ceremony in a field of soon-to-be sunflowers. The company's new initiative looks to sunflowers as a source of biofuel, estimating that the plants have the potential to produce about 34,500 gallons of fuel per year. Though the terrain is harsh, Pacific Biodiesel's research shows that their sunflowers actually produce more oil in a windy environment. The company has tested up to ten varieties of sunflowers, strategically selecting the two they believe will have the most potential in Maui soil. The estimated date for the sunflowers' first harvest is on Earth Day, April 22nd, and will hopefully be celebrated with an event open to the community.

The initiative utilizes 115 acres of farmland near the intersection of the Kuihelani and Honoapiilani highways that was previously used to grow sugar cane, but would no longer be in use after the end of 2016. Sunflowers are actually just the first crop in a series of crop rotations that the company has planned out. Other crops include safflower, camelina, hemp, and non-genetically engineered soybeans. They are all short-term crops that harvest in 100 days or less. The crops from the sunflowers harvest will undergo processing at Pacific Biodiesel's Big Island refinery, where the company states that one gallon of virgin sunflower oil yields 0.95 gallons of biodiesel.

According to Pacific Biodiesel, this biofuel project is the largest in the state of Hawaii, as well as the only project running completely on renewable fuel. The company hopes to farm as naturally as possible, using rain instead of irrigation systems and partnering with Maui EKO Systems to provide natural fertilizer instead of manufactured fertilizer. Pacific Biodiesel's commitment to farm 265 acres will last at least five to ten years.

http://www.biodieselmagazine.com/articles/2360112/pacific-biodiesel-launches-hawaiis-largest-biofuel-crop-project
http://mauitime.com/news/science-and-environment/pacific-biodiesel-begins-farming-sunflower-biofuel-crop/
http://www.mauinews.com/news/local-news/2017/02/something-in-the-wind/

Pacific Northwest Company Researches Poplar as Source of Biofuel

by Nina Lee

The Advanced Hardwood Biofuels Northwest study focuses on a crop that may not look like anything out of the ordinary, but holds potential for being a new source of biofuel in Oregon. The study is the combined effort of multiple academic and industrial partners, including Oregon State University, and has four sites in the Pacific Northwest. One of these sites is an 80 acre plot of federally-funded land outside the town of Jefferson, Oregon. There are over 115,000 trees at this site, many of which were being harvested as of the end of February 2017.

While the current focus for the study is to convert the poplars into jet fuel, they can also be converted into gasoline, diesel, ethanol, and other chemicals. One acre of poplar trees yields about 2,000 gallons of ethanol every three years. Since the project was technically started during the recession when gas prices were considerably higher, Advanced Hardwood Biofuels Northwest needed to adjust their goals as the economic climate changed. Now, gas prices would need to be double the amount they currently are in order for Poplar biofuel to become a major player in the biofuel market.

Oregon State University is currently collaborating on this project by running a program that educates middle school and high school students about biofuel and the potential of poplar trees in their area, providing important knowledge for when biofuel becomes a more economically sound option.

The reason that Poplar has garnered more attention as of late is due to new biofuel rules that were proposed at the end of 2016. The original Renewable Fuel Standard was passed in 2007 in order to reduce greenhouse gas emissions and encourage the production of cellulosic ethanol, and the EPA plans to make changes in order to allow cellulosic ethanol to be made from poplar and willow trees.

http://democratherald.com/news/local/poplar-trees-being-used-in-biofuel-study/article_3a13aa41-e153-5df7-8344-9c914da0e124.html
http://fortune.com/2016/10/03/willow-trees-epa-biofuel-rules/
https://cleantechnica.com/2016/10/05/us-to-opec-no-worries-well-make-biofuel-from-trees/

Biomass Power

Factors Influencing Traditional Biomass Use in Cooking

by Emily Audet

About a third of the world's population uses traditional biomass—like firewood, charcoal, agricultural products, and animal excrement—to cook food with.

Burning biomass indoors creates large amounts of air pollution, which in turn, has severe negative health effects on individuals within the household. Many initiatives in the last 20 years have tried to shift households from using biomass to cleaner fuel types; however, this goal has largely not been achieved, and critics claim that the failure of these initiatives lies in their inattention to context-specific household needs, focusing instead on the logical reasons for switching to cleaner fuel.

In 2012, the Stockholm Environment Institute (SEI) conducted household surveys in India, where about a third of traditional biomass users reside, in order to improve understanding of factors that encourage households to transition to cleaner fuels. The study found a diversity of variables that influence household fuel use decisions, many of which were rooted in individual and cultural preferences. SEI found that households would consider switching to cleaner fuels if these methods used less fuel, decreased smoke, decreased the time it takes to cook and collect fuel, and were more mobile, among others. They also found a multitude of reasons why households value their current cooking methods, including a preference for the taste of food cooked with biomass and the convenience of not having to buy the stove or fuel. All of the study's interviewees said they would consider buying a cleaner stove if the stove would use less fuel. Many households stated that they would be willing to pay between 100 to 250 rupees, about $1.50 to $4, for a new stove.

In a March 2017 paper, Sandra Baquié and Johannes Urpelainen of Columbia University test the hypothesis that people using cleaner fuels, like liquefied petroleum gas (LPG), are more content with their fuel than those using biomass. Baquié and Urpelainen sampled households in rural India and compared those with access to cleaner fuels to those without, to assess household satisfaction with their cooking situation. They found that access to LPG was most correlated with satisfaction. Based on these findings, the authors postulate that the cost of cleaner fuels is the factor deterring rural households from transitioning to them, but they recommend further studies to measure this relationship.

[http://www.theenergycollective.com/jurpelai/2401398/value-modern-cooking-fuels-evidence-subjective-satisfaction-rural-india; international.org/mediamanager/documents/Publications/Atmospheric/sei-wp-2012-03-cookstoves.pdf].

Lambe, Fiona, and Aaron Atteridge. "Putting the Cook Before the Stove: A User-Centered Approach to Understanding Household Energy Decision-Making." May 2012. Web. 4 Apr. 2017. https://www.sei-international.org/mediamanager/documents/Publications/Atmospheric/sei-wp-2012-03-cookstoves.pdf

Upelainen, Johannes, and Sandra Baquié. "The Value of Modern Cooking Fuels: Evidence on Subjective Satisfaction from Rural India." The Energy Collective. N.p., 30 Mar. 2017. Web. 2 Apr. 2017. http://www.theenergycollective.com/jurpelai/2401398/value-modern-cooking-fuels-evidence-subjective-satisfaction-rural-india

Waste-to-Energy Industry Growing in Popularity Across the Developing World

by Lauren Bollinger

This year has seen a significant increase in the development of the waste-to-energy (WtE) industry across the developing world. Praised as a solution to overflowing landfills, WtE technology involves the conversion of waste (often through incineration, chemical conversion, or other methods) to electricity, heat, or fuel. In Cape Town, South Africa, a WtE conversion plant was opened in mid-January, while officials from United Arab Emirates announced plans for the development of two such facilities. Both Indonesia and India are in the process of constructing WtE plants, amid bureaucratic and legal barriers, while the Philippines has seen recent debate over WtE technology, which is currently banned in the country.

These countries' interest in WtE technology is no coincidence; all share similar problems concerning waste management and energy development. Namely, these developing countries face the problems of increasing urban waste and overflowing landfills and significant concerns over air and water quality, making the WtE technology an appealing and relatively environmentally-friendly option.

Nonetheless, many leading energy scientists remain skeptical of the adoption of WtE technology. As Forbes reports, Jorge Emmanuel, a professor at Silliman University in the Philippines and leading voice on renewable energy in the country, has criticized recent moves to develop WtE technology, citing concerns over the release of harmful toxins in the atmosphere and the country's lack of technical capacity to moderate such toxins. Instead, Emmanuel argues, the country should invest in clean energy, such as solar, wind, and geothermal industries, which pose less threat to the environment and population.

These debates over waste management and WtE technology are especially relevant in such nations, as Third World countries like the

Philippines are increasingly becoming the First World's dumping grounds.

Despite its premise as a solution to urban waste and energy crises, WtE technology still remains a controversial option in developing countries.

From Cape Town To Jakarta, Cities Are Choosing WtE To Fix Their Landfill Problem
http://www.forbes.com/sites/nishthachugh/2017/01/31/from-cape-town-to-jakarta-
        cities-are-choosing-wte-to-fix-their-landfill-problem/2/#368736895ac6
Waste-to-energy plant opens in Cape Town
http://www.news24.com/SouthAfrica/News/waste-to-energy-plant-opens-in-cape-town-
        20170125
Waste-to-energy technologies in PH? 'Go zero waste instead'
http://www.rappler.com/science-nature/environment/159495-zero-waste-month-waste-
        energy-technologies

Chicken Manure into Energy: Maryland Project Key to Cut Pollution

by Lauren Bollinger

An Irish agri-tech company has found an innovative solution to the longstanding pollution concerns faced by Maryland's chicken industry. The firm, Biomass Heating Solutions Limited (BHSL), has recently partnered with the Maryland Department of Agriculture to pilot a chicken manure-to-energy conversion project. Based on 112-acre chicken farm in Rhodesdale, Maryland, the project went live in mid-February following a $3 million grant by the Maryland office.

The project utilizes a process called fluidized bed combustion, which works by first heating a bed of sand inside a fuel combustion chamber to 600 degrees Fahrenheit, feeding manure into the chamber, and then raising the temperature to 1,600 degrees. This process produces hot gases, which can then be used to provide heating for future batches of chicks or for electricity generation which can be used to cool sheds in the hotter summer months.

The technology serves dual purposes: helping divert up to 10 tons of manure a day—that would otherwise pollute local ecosystems—and providing the chickenhouses with clean, renewable energy.

Though the project only oversees a fraction (160,000 out of 300 million) chickens the state breeds per year, it has engendered high hopes for significantly reducing the impact of the local poultry industry on the nearby Chesapeake Bay.

Maryland is one of the country's top producers of chickens for meat consumption—in 2013, the state produced 305 million chickens, making it eighth in the nation for chicken production. The state's Eastern Shore is an especially productive area for the poultry industry —chickens account for the largest sector of the Eastern Shore's agricultural industry.

Though manure-to-energy based projects are not new, this project shows promise as a new solution to longstanding problems facing the Chesapeake Bay area, as a unique solution to both pollution issues and energy demands.

Maryland farm turning manure into energy
http://www.delmarvanow.com/story/news/local/maryland/2017/02/14/chicken-manure-
    renewable-energy/97895002/
Irish agri-tech company launches poultry manure pilot in US
http://www.farmersjournal.ie/irish-agri-tech-company-launches-poultry-manure-pilot-in-
    us-255331

Waste to Energy

by Dominique Curtis

The Maryland Department of Agriculture and Biomass Heating Solutions Limited have committed approximately $3 million towards manure-to-energy technology that they hope will reduce the pollution and energy impact of Murphy's chickens. The technology will generate energy from the animal manure, reduce on a farm waste streams, and repurpose the manure.

According to Chavez (2017), Biomass Heating Solutions Limited uses a process called fluidized bed combustion to generate energy from the manure. The process works by heating a bed of sand inside a fuel combustion chamber until it starts bubbling at 600 degrees Fahrenheit. After this the manure is put into the chamber and the temperature is raised to 1,600 degrees Fahrenheit. The excess heat is used to boil water and heat the chicken houses.

Bob Murphy from Double Trouble Farms believes that this technology will be a means to cleaning the Chesapeake bay, and saving energy, while preserving the chicken economy. He says that 90% of the poultry litter produced will have alternative uses because of energy production (2017).

Murphy says he hasn't heard of any opposition from the locals and that other people in the poultry industry have given their approval of the project. There is chatter about bringing this technology to Bellview farms as well.

This manure to energy project will reduce environmental impact, lower energy costs, improve animal welfare, and create additional revenue.The goal is to use Double Trouble Farms as a test project and example for other farms to get them to try this new process of managing manure in a way that is healthy for the chicken and the environment.

Chavez, Jack. "Maryland Farm Turning Manure into Energy." Delmarva Daily Times. N.p.,
    14 Feb. 2017. Web. 21 Feb. 2017.
Energy Management. "Double Trouble Farm Installs Manure-to-energy
    Technology."Refrigerated Frozen Food RSS. BNP Media, 16 Feb. 2017. Web. 21
    Feb. 2017.

Biomass to Clean Hydrogen

by Dominique Curtis

Scientists at the University of Cambridge have developed a way to use natural light to generate hydrogen from biomass. The technology was developed in the University of Cambridge Laboratory for

Sustainable Chemistry. The scientists are generating a fuel that is both sustainable and cheap to produce by using solar power. This is useful for our society because it's turning our waste into energy.

According to the scientists "Biomass has been a source of heat and energy since the beginning of recorded history" (University Of Cambridge, 2017). They explain that Lignocellulose is the main component of plant biomass and before now it could only be converted into hydrogen through a gasification process that uses high temperatures to decompose the biomass. The new technology uses photocatalytic conversion process. There is a solution created using alkaline water and catalytic nanoparticles, that absorbs light and then converts the biomass into gaseous hydrogen that can be collected; the biomass doesn't have to be processed beforehand. When testing the process the scientists used different types of biomass such as wood, paper, and leaves.

The scientists see this as an alternative to high temperature gasification and a new way of producing renewable hydrogen. In the future they hope to create large scale plants and find off-grid applications.

University of Cambridge. "Scientists Harness Solar Power to Produce Clean Hydrogen from Biomass." Science Daily. Science Daily, 13 Mar. 2017. Web. 19 Mar. 2017.

Wakerley, David W., Moritz F. Kuehnel, Katherine L. Orchard, Khoa H. Ly, Timothy E. Rosser, and Erwin Reisner. "Solar-driven Reforming of Lignocellulose to H2 with a CdS/CdOx Photocatalyst." Nature News. Nature Publishing Group, 13 Mar. 2017. Web. 18 Mar. 2017.

The Looming Threat of Climate Change on Peatlands

by Ethan Fukuto

In January 2017, Simon Lewis of the University of Leeds published new findings mapping 55,000 square miles of peatlands in the Cuvette Centrale depression in the Congo Basin of Central Africa. Lewis et al.'s findings chart the Cuvette Centrale as the largest peatlands in the tropics, containing around 30 percent of the worldwide total of carbon soil in tropical peatlands. Carbon-rich peat, perhaps best known as an ingredient in whisky production, is a soil formed by decomposing organic matter. Peatlands are found primarily in northern regions such as Canada and Europe, though tropical peatlands, such as the Cuvette Centrale, pose a greater risk to global climate issues. As these regions dry due to climate change and human land-use, their susceptibility to fires increases the risk of a massive output of carbon into the atmosphere. A 2006 study on soil carbon and climate change by Eric Davidson and Ivan Janssenns called for a broadening of scope in the study of temperature sensitivity to include areas such as peatlands. They describe these areas, along with wetlands and permafrost soils, as the "most obvious environments" affected by "climatic disruption", and thus necessitate further study and scrutiny. In 2015, Indonesia endured intense fires across peatlands as a result of the palm oil industry's slash and burn farming techniques. Indonesian peatlands likewise released between 0.6 to 0.8 petagrams (10^{15} grams) of carbon alone during the 1997 El Niño, accounting for 10% of anthropogenic emissions. Lewis et al. estimate that the Cuvette Centrale peatland contains around 30.6 petagrams of carbon belowground. Given its tropical environment, the recently-discovered peatlands are at risk of fires and intense carbon emissions due to potential changes in land-use. Current legal protection of the region is scant: the region is relatively inaccessible to humans; however this does not preclude the potential for future development. Lewis et al.'s study is the first step in preserving and protecting the region. Given the historical example of Indonesia's peatlands, this study calls for greater attention to the Cuvette Centrale in planning and government initiatives.

Davidson, Eric A., and Ivan A. Janssens. "Temperature sensitivity of soil carbon decomposition and feedbacks to climate change." Nature 440, no. 7081 (2006): 165-173. (http://www.nature.com.ccl.idm.oclc.org/nature/journal/v440/n7081/full/natu re04514.html).

Fountain, Henry. "As Peat Bogs Burn, a Climate Threat Rises". New York Times, (8 August 2016). (https://www.nytimes.com/2016/08/09/science/climate-change-carbon-bogs-peat.html).

Fountain, Henry. "Scientists Map Vast Peat Swamps, a Storehouse of Carbon, in Central Africa". New York Times, (11 January 2017). (https://www.nytimes.com/2017/01/11/science/peat-swamp-congo-global-warming.html).

Lewis, Simon, Dargie, Greta C., Ian T. Lawson, Edward TA Mitchard, Susan E. Page, Yannick E. Bocko, and Suspense A. Ifo. "Age, extent and carbon storage of the central Congo Basin peatland complex." Nature (2017). (http://www.nature.com/nature/journal/vaop/ncurrent/full/nature21048.html).

Brown Gold

by Byron R. Núñez

Biomass Heating Solutions Limited (BHSL), an Irish agriculture and technology company, has developed Fluidized Bed Combustion (FBC) technology to reliably convert poultry manure into heat and electricity for use on poultry farms. How does it work? Well, poultry manure is transferred to a bio-secure building where it is combusted. The heat generated by the process can be vented into the chicken houses to keep them warm, while the electricity created can be used to meet the energy demands of the farm.

This technology has mainly been used in Ireland and the United Kingdom. However, a Maryland Department of Agriculture grant of $1 million and a private $2 million investment by BHSL in Dorchester, Maryland are funding the first six projects in the United States. These pilot projects are projected to reduce the environmental impact of manure, lower energy costs for heating, improve animal health and create additional revenue for the farms from the sale of excess electricity as well as from the fertilizer by-product, which is an ash that can be sold as a nonpolluting fertilizer.

The grant from the state comes from the Maryland's Animal Waste Technology Fund, which "provides seed funding to companies that demonstrate innovative technologies to manager repurpose manure resources." Agriculture is Maryland's number one industry, which explains why the state is interested in this revolutionary technology.

FBC is hailed as a solution to poultry waste as it generates energy from animal manure, reduces on-farm waste streams, and repurposes manure by creating marketable fertilizer and other products and by-products. This technology's ability to create a nonpolluting fertilizer will have a tremendous environmental impact on Maryland's Eastern Shore. Dorchester farms, for example, produce 3,650 tons of manure annually, which is trucked to other farms as fertilizer. This practice is challenged by phosphorus management pollution reduction regulations since chicken manure has high phosphorus content. One of BHSL's goals is to turn a potential pollutant, poultry manure, into a valuable source of energy.

Dorchester Star
http://www.myeasternshoremd.com/dorchester_star/news/article_8a848e9c-5a0c-53de-aa6f-9c068fec0c89.html
TAGS: Agri-tech, Biomass Heating Solutions Limited, Fluidized Bed Combustion, poultry manure, Maryland Department of Agriculture, fertilizer, by-products, pollutant.
info@bhsl.com, https://www.facebook.com/BHSL.ie/, https://www.linkedin.com/company/bhsl, @BHSLTeam

Municipal Solid Waste: To the Landfill or the Incinerator?

by Nadja Redmond

A global phenomenon is slowly beginning to pick up traction and conversation in the United States: energy recovery through use of waste to energy facilities. WtE, the waste management process that involves generating electricity and/or heat from waste through combustion, is already widely used in Europe. By 2014, Europe had 452 such facilities, and compared to the United States' 71, it is no secret there is an ongoing debate on whether WtE facilities are effective or hazardous for the environment and for the communities they inhabit. When the country produces over 250 million tons of municipal solid waste a year, alternative routes of waste management and energy recovery that utilizes that waste that have proved effective overseas are worth considering.

Advocates and some sustainability researchers argue waste to energy is the best process for the environment to eliminate waste, because of the long-term benefits of less landfill use and proper disposal of waste materials that are not recyclable. Opponents argue the emissions from WtE plants only add to the pollution already present in the atmosphere. A recent fire at a plant in Maryland at a Covanta operated facility, which maintains that the conditions of the facility did not cause the fire, also causes speculations about the process among opponents. In

Baltimore specifically, these plants are blamed for failure to meet federal ground ozone level standards. According to the EPA, these plants actually help reduce greenhouse gas emissions as compared to landfills, which only proves advocates and opponents alike need to learn more about the benefits of these types of facilities. Covanta chief officer Paul Gilman stresses the importance of the facility in an ongoing investigation. For example, its partnerships with American Airlines and Subaru allow the corporations to contribute zero waste to landfills because of the company's initiatives with hard to dispose of materials. The success of these initiatives is being recognized slowly around the country, with Clean Harbors completing an expansion of a waste incineration plant in South Arkansas recently. This is the first construction in many years since the debate on renewable energy came to light.

[http://www.environmentalleader.com/2017/01/waste-to-energy-facilities-under-fire/].
[https://www.eia.gov/todayinenergy/detail.php?id=25732],
[http://www.cewep.eu/information/data/studies/m_1488],
EPA - https://www.epa.gov/smm/advancing-sustainable-materials-management-facts-and-figures

The Solution to Lack of Light in the Rainforest: "Plant Lamps"

by Yerika Reyes

Researchers at the Universidad de Ingeniería y Tecnología (UTEC) have developed a method for capturing the electricity released from plants, specifically the soil. They collect a microorganism called, geobacter, a genus of bacteria that live in the soil. Nutrients in plants encounter the geobacters in the dirt, and that process releases electrons that electrodes in the dirt can capture. A grid of these electrodes can transfer the electrons into a standard battery. UTEC has partnered with global ad agency Foote, Cone & Belding, or FCB to produce 10 prototypes and distribute them to houses in the rainforest village of Nuevo Saposoa. Each contains an electrode grid buried in dirt, in which a single plant grows. The grid connects to a battery, which powers a large LED lamp attached to an adjustable arm on the outside of the box. UTEC coined this invention "plant lamps."

The hope is that if the initial testing period is successful, their appeal isn't going to be limited to rainforest communities. Hopefully, these plants can become part of people's houseplant collection which will reduce the electrical bill and add some greenery to their wallets and their home.

UTEC's researchers are not the first to make use of geobacter. In 2009, Time named the "electric microbe" one of its 50 best inventions of the year. In addition to power generation, geobacter have also garnered attention for their ability to metabolize pollution like radioactive material. What sort of power could an entire garden generate? Is there a way to combine pollution-tolerant plants with the electric grid and bacteria? Is it possible to have a grove of trees help reduce soil pollution and provide power? UTEC's plant lamps are the perfect model of using technology to bring people into more mutually beneficial relationships with nature and to solve local problems. It is not only remarkable, it is symbiosis.

http://www.bbc.com/news/business-18262217
http://schaechter.asmblog.org/schaechter/2011/03/geobacter-microbial-superhero.html
http://content.time.com/time/specials/packages/article/0,28804,1934027_1934003_1933
965,00.html

Mexican Tequila Companies Turning Agave to Fuel

by Sara R. Roschdi

Tequila companies in Mexico are finding ways to turn their waste into renewable energy sources in an effort to reduce the cost of energy used in the production of Tequila. The Jose Cuervo producer, Herradura, has started to dry out their agave waste in order to generate energy to be used in steam plants. They dry out approximately 150 tonnes of agave waste and use it as biomass to fuel the boilers. This process squeezes out the water from the fibrous agave plants and uses it as fuel. The company is struggling to innovate and gain access to machinery to dry and process the agave waste. This method currently produces 20% of the company's energy and has the potential to supply 30% of the energy of the plant. The Tequila Regulatory Council reports that in 2015 Mexico produced approximately 230 million liters of tequila and exported about 180 million liters University of Adelaide reports that agave is "predicted to yield between 4000 and 15,000 liters of ethanol per hectare per year (Bourne, 2015). Herradura announced a new partnership with Ford to put these leftover agave plants to use building automobiles. They are exploring turning fibrous bits of the agave plants into bioplastic that can be used to build certain car parts. This is the first partnership of its kind and an introduction to the use of agave in the automobile industry. The bioplastics made by agave are anticipated to improve the fuel efficiency of cars by being a lightweight alternative to petroleum-based plastics. Agave fibers are anticipated to be stronger and have better quality mechanical properties than previously used less environmentally friendly models. The Ford Company and Herradura Tequila Company are teaming up to reduce greenhouse emissions, eliminate waste and produce higher quality sustainable vehicles.

https://www.washingtonpost.com/news/energy-environment/wp/2016/07/22/why-ford-
is-making-car-parts-out-of-carbon-dioxide-and-plants/?utm_term=.baea7f54d1df
http://www.reuters.com/article/us-mexico-beverages-tequila-energy-idUSKBN1631CE.
http://www.telesurtv.net/english/news/Mexico-Tequila-Maker-Aspires-to-Turn-Agave-
Waste-into-Fuel-20170224-0004.html

Bio-Bean: Using Coffee Grounds as Fuel

by Chloe Soltis

In 2013, Arthur Kay founded Bio-bean, a London-based start-up that is developing technology to use coffee grounds as a fuel source. When Kay was an architect student, he learned that coffee shops produce hundreds of thousands of pounds of coffee grounds as waste. Then, the shops pay for the grounds to be disposed in landfills (Huffington Post). Upon learning this, Kay began researching alternate methods for recycling the coffee grounds. Ultimately, he started Bio-bean and developed a procedure for processing the grounds into biomass pellets. When densely packed into pellets, coffee has a significantly higher calorific value than wood, which means that it releases more energy when it burns. The pellets primarily are used to heat buildings and power industrial boilers (Huffington Post).

Bio-bean has also developed a procedure for processing the coffee grounds into coffee logs. The logs, or briquettes of compressed grounds, can be used in chimneys, open fires, and even BBQs. The logs are a 100% carbon-neutral biofuel and burn at a higher temperature for a longer period of time than wood logs (Bio-bean). In addition, coffee logs are half the price of wood logs (Telegraph). Bio-bean is currently researching how coffee grounds can be used for biodiesel and eventually wants to learn how extracts from the grounds could be used in biochemicals.

Bio-bean currently operates the world's first coffee grounds recycling plant in Alconbury Weald, England and a factory that produces their products in Cambridgeshire, England. However, Kay is actively looking to expand the company since Bio-bean's technology and business model can be implemented in any country that drinks coffee (Telegraph). Bio-bean turned 10% of England's coffee waste, 50,000 tons of coffee grounds, into energy in the past year, which is enough to power 15,000 homes (Huffington Post). Kay plans for Bio-bean to increase its processing capacity to 250,000 tons over the next few years. Bio-bean's goal is to have both people and their cities powered by coffee.

Bio-bean. http://www.bio-bean.com/collection/.

Burn-Callander, Rebecca. This Fuel Made From Old Coffee Will Launch in the Summer, at Half the Price of Wood. The Telegraph. Feb 20, 2017: http://www.telegraph.co.uk/business/2016/04/11/fuel-made-from-old-coffee-to-launch-this-summer-at-half-price-of/.

Ridley, Louise. Bio-Bean Entrepreneur Arthur Kay Is Turning Coffee Into Fuel And Wants To Power All Of London's Buses. Huffington Post. Feb 20, 2017: http://www.huffingtonpost.co.uk/entry/bio-bean-coffee-fuel-waste-arthur-kay_uk_57109444e4b0dc55ceea465b.

Geothermal Power

Volcanic Geothermal Energy: Iceland's Secret to Clean Energy

by Sagarika Gami

After seven months of anticipation, the Iceland Deep Drilling Project (IDDP) completed drilling into the core of a volcano, which has been dormant for more than 700 years, in the Reykjanes Peninsula. It took 168 days of drilling until the well was completed on January 25, 2017, reaching a record depth of 4.8 kilometers. At this depth, the well does not reach the magma chamber, but does penetrate the rock around it, which reaches a temperature of 427 degrees Celsius, it may get even hotter as the well widens. This project harnesses geothermal energy, a major source of energy in Iceland. The National Energy Authority of Iceland shows that 25% of the country's electricity is generated via the Earth's heat and 90% of Icelandic households are heated via geothermal energy.

Geothermal energy uses the heat under the Earth's surface to generate energy, usually conducted by steam from geysers or by drawing water from high-pressure depths of Earth that is then used to drive electric turbines. Volcanic geothermal energy works a little differently, the heat comes from the meeting of molten rock and water, creating "supercritical" water that is neither liquid nor gas. In this form, water holds more energy than either. Supercritical water brings enough energy to generate up to 10 times the power output of other geothermal sources.

The next stage of the project will be to pump cold water into the well to open it up and then wait for the well to heat up again; it is expected to exceed temperatures of 500 degrees Celsius. The research is funded by energy companies (HS Orka, Statoil, Landsvirkjun and Orkuveita Reykjavíkur), Orkustofnun, the International Continental Scientific Drilling Program, the National Science Foundation in the US, and EU Horizon 2020. Research will continue into 2018 to understand how the volcano's thermal energy can be used as another alternative source. Iceland remains one of the cleanest energy countries at 100%, and is looking to keep it that way.

Livescience.com (http://www.livescience.com/57833-scientists-drill-volcano-core-geothermal-energy.html)

Electrek.co (https://electrek.co/2017/02/10/electrek-green-energy-brief-austin-building-code-goes-solar/)

Bbc.com (http://www.bbc.com/news/science-environment-38833023)

Tidal and Ocean Energy

Energy Potential of the Ocean

by Genevieve Kules

According to William Steel, marine energy is one of the most underutilized and underdeveloped forms of renewable energy harvesting. There are numerous types of potential energies within the ocean and waters on this planet, such as wave energy, tidal energy, and osmotic power (which is created through varying levels of salinity in the water).

Wave and tidal energy both harness the movement of the ocean to capture energy, although the method for each is a little different. Wave energy is usually gathered from a floating device anchored to the bottom of the ocean floor. Tidal energy is often harvested through underwater turbines; when the current moves they spin and create electricity, like a wind turbine. My concern, along with many others, is the safety of marine wildlife. Fish and other animals could get and have gotten severely hurt and died from these turbines.

Steel discusses a few projects underway in Europe that are pioneering parts of the marine energy field. He conveys a statement by Ocean Energy Europe (OEE) that by 2050 they could be producing over 300 GW of marine energy generated primarily by tidal turbines. OEE states that ten marine energy devices have been deployed in the last three years, in comparison to the three that had been deployed before that. Steel recognizes how far behind this technology is, but believes strongly in its potential. He reminds the readers that about 72% of the earth's surface is covered by water, and recognizes that therefore it is one of the most underused forms of energy production.

Steel, William. "On Advances in the Marine Energy Industry." Blog post. Phlebas. WordPress, 15 Apr. 2017. Web. 24 Apr. 2017. <https://phlebas.co/2017/04/15/on-advances-in-the-marine-energy-industry/>.

"Ocean Energy Project Spotlight." (n.d.): n. pag. Ocean Energy Europe, Mar. 2017. Web. 24 Apr. 2017. <https://www.oceanenergy-europe.eu/images/Documents/Publications/170228-Ocean-energy-spotlight-final.pdf>.

Carnegie Clean Energy. Carnegie Wave Energy Limited, 2015. Web. 24 Apr. 2017. <http://carnegiewave.com/>.

The Wave Energy Prize

by Nadja Redmond

Scientists and NASA are teaming up to decipher ways to best capture the kinetic energy produced by waves. Oceans have a vast potential to become a new renewable energy source because of this constant movement of water across the surface. However, the ability to harness power from waves has proven difficult. Kinetic energy collected from the ocean could be used to create electricity, but the mechanical wave energy converter (WEC) that is necessary for this transformation is a complicated mechanism and is still being perfected.

More than 50% of the US population lives along the coasts, which makes waves an ideal renewable energy source. While many European countries have been exploring wave energy since the 1970s, the United States have done much less so. Recently, the U.S. Department of Energy introduced the Wave Energy Prize, a public prize designed to award achievements and diversity in wave energy converter technology. The long-term goal is to reduce the cost per kilowatt hour of wave energy. Recently, the cost has reduced from $0.90 per kilowatt hour in 2015 to $0.66 per kilowatt hour earlier this year, with an objective of $0.17 per kilowatt hour by 2030.

Other goals of the project are the development of water energy converters that can withstand varying weather and saltwater corrosion. The most common model currently is a buoy tethered the seafloor which transfers energy through an underwater cable. A successful power transfer relies on resonance in the system, or how synced the natural frequencies of the device and the waves are. Rob Berry, a NASA project manager who deals with resonance issues in space, believes methods similar to those used to control resonance in space can be used in the water energy converters. His ideas involve using sea water to change the resonant conditions of the buoy, so that no matter what external vibration occurs, there isn't a huge interruption of the resonant system. Berry said: "We have the ability to ... control the way the structure behaves in water. This (new) fluid/structure coupling ability could lead to a totally different design and significantly reform what existing designs can do. There are always those innovators out there and finding those people is key."

[https://www.pastemagazine.com/articles/2017/03/making-energy-with-waves-and-nasa.html]

Tidal and Wind Energy Companies Share a Power Grid to Provide Reliable, Renewable Energy

by Mary-Catherine Riley

Atlantis Resources partnered with Lockend Wind Energy to spearhead the world's largest grid connection of any commercial tidal project. This initiative is thought to be the first combination of electricity to power an existing grid. Currently, MeyGen is in the first phase of construction, installing 86 turbines to generate 86 megawatts (MW). However, the project has room for growth. Atlantis hopes to expand the facility's capability to power 175,000 homes using 269 turbines producing almost 400 MW (3,4). The glaring downside is the cost. Funds for the first stage of the MeyGen project are £51million ($82m). Moreover, while the power of strong currents in the Pentland Firth in northern Scotland makes it an ideal location for tidal generation, the area's harsh storm and wave conditions could destroy the turbines. Lastly, the grid connection is limited until further expansion occurs in future years due to limited grid capacity.

However, this project serves as a great positive for the future of energy. This intermittent use of wind and water energy could eliminate the need to employ nuclear or coal power plants while still reaching maximum energy capacity. Moreover, European environmental corporations predict that Scotland has 25% of the European Union's offshore tidal and wind energy potential, so this expansion begins to actualize this potential source of energy. This project is essential in testing how reliable and effective a commercial scale water turbine farm is to power a grid.

Moreover, MeyGen is dedicated to boost the local economy by providing highly skilled technical and labor jobs to retain the Scottish workforce. Moreover, this project contributes to the European Union's larger goal of saving 20% of the projected consumption of energy in 2020 in 2020

(https://eandt.theiet.org/content/articles/2016/11/meygen-tidal-power-project-poised-to-feed-scottish-highland-electricity-grid/).

(https://ec.europa.eu/energy/en/topics/energy-efficiency).

(http://www.meygen.com/the-project/meygen-news/
(https://eandt.theiet.org/content/articles/2016/11/meygen-tidal-power-project-poised-to-feed-scottish-highland-electricity-grid/).

"Energy Efficiency - Energy - European Commission." Energy. European Commission, n.d. Web. 24 Jan. 2017.

"Spring 2016 Project Update." MeyGen. MeyGen Limited, n.d. Web. 24 Jan. 2017.

Thomson, Corrina. "MeyGen Tidal Power Project Poised to Feed Scottish Highland Electricity Grid." Engineering and Technology. The Institution of Engineering and Technology, 08 Nov. 2016. Web. 24 Jan. 2017.

Harvesting Energy from Ocean Waves with eForcis

by Chloe Soltis

In 2012, Héctor Martín and Rubén Carballo founded Smalle Technologies, a Barcelona-based company that designs and assembles products that harvest clean energy. Most recently, the start-up has been developing and testing eForcis, a device that harvests kinetic energy from the oscillation of ocean waves. The purpose of eForcis is to supply electricity to marine devices such as moored buoys that are off the grid. Moored buoys float in the ocean and can measure variables such as wave height, wind speed, air temperature, and barometric pressure (Smalle Technologies).

Solar panels, batteries, and small wind turbines have traditionally been used to electrically power buoys. Unfortunately, these energy forms can either contaminate the ocean or are high maintenance due to the harsh marine conditions. eForcis is different because the entire system is enclosed in a sealed box, which protects it from salt water or wind corrosion, and does not need sun or wind to operate (Crowdfund). The eForcis can generate electricity from the tilting motion of both large and small waves and needs little maintenance due to its enclosed design, which can dramatically decrease expenses for the buoys' operators. The operators can also easily install the eForcis onto new or old buoys; the system is adaptable and can replace previous energy systems. In addition, almost all the materials used to manufacture the eForcis can either be reused or repurposed after its lifetime (Smalle Technologies).

Over the past year, Smalle Technologies has been testing eForcis in different locations off the coast of Spain. Testing has been going well so far and the company reports that it wants to sell and install 4,000 eForcis units over the next 5 years (Smalle Technologies). Smalle Technologies is also developing another electric generator called the BeForcis, which will be installed on buoys that are involved in fish farming.

Alois, JD. Energy Startup Smalle Technologies Raises €240,550 on Crowdcube Spain. Crowdfund Insider. Feb 13, 2017: https://www.crowdfundinsider.com/2017/01/94972-energy-startup-smalle-technologies-raises-e240550-crowdcube-spain/.
Smalle Technologies. https://twitter.com/smalletec.

Kinetic Energy

Turning the Kinetic Energy of Everyday Movements into Light

by Nina Lee

What would happen if we thought of energy not as a resource to be saved up, but as a resource to be constantly created and used? Uncharted Play, the brainchild of Jessica O. Matthews, is a company that challenges the way our society values energy. Matthews is a Nigerian-American Harvard graduate, inventor, and CEO who was inspired to create alternative energy sources after an experience with her family in Nigeria. During her aunt's wedding, there was a power outage- a very common occurrence- and diesel generators had to be used to supply energy. The generators were emitting toxic fumes that everybody but Matthews seemed to be used to. When she later returned to the United States and continued her education, she wanted to create a cleaner alternative to the energy sources her community in Nigeria were utilizing.

The first product that Uncharted Play launched was called the SOCCKET, a soccer ball that generates energy from the motion of the ball being kicked around and harnesses that energy using a micro-generator and a rechargeable lithium ion battery. Uncharted Play has also released the PULSE, a jump rope that utilizes the same concept as the SOCCKET, taking the rotational energy from each turn of the rope and using a micro-generator to charge a lithium ion battery.

While the company ended up experiencing some manufacturing difficulties, they ended up coming back stronger than ever and changed the trajectory of their mission. Instead of focusing on the quality of the products they were putting out, the company's new objective is to instead perfect the M.O.R.E. (motion-based off-grid renewable energy) technology and partner with existing companies to install M.O.R.E. technology into their partner's products. M.O.R.E. technologies can harness energy from modular rotational, linear compression, and triboelectric motions. That, in combination with specialized circuitry that gathers energy from various micro-generator systems, can power batteries, sensors, WiFi, and light. Matthews hopes to incorporate M.O.R.E. technology into more public, larger scale areas, such as parks, gyms, and subway turnstiles.

Harvesting Energy from the Dance Floor

by Byron R. Núñez

Club Surya in London is one of an increasing number of "eco" nightclubs that harnesses the power of dancers. The dance floor is fitted with springs and a series of power generating blocks, making the floor "bouncy." As dancers move up and down, the blocks are squashed and the energy created is fed into nearby batteries. The batteries are constantly recharged by the movement of the floor and are used to power the nightclub's sound and lighting systems. This process is known as piezoelectricity, which refers to the electric polarity resulting from the application of mechanical stress.

Andrew Charalambous, owner and developer of Club Surya, estimates that the dance floor supplies 60% of the club's energy needs. To meet the other 40%, Charalambous has installed a wind turbine and solar panels on the roof of the club. The owner hopes to inspire young people to combat global warming by showing them how renewal sources of energy can sustainably power almost anything.

In Rotterdam, Netherlands, Club Watt also houses an energy harvesting dance floor which generates power for the nightclub's lighting system, with the average dancer making around 20 watts of electricity. One of the first nightclubs to use piezoelectricity was the Sustainable Dance Club in the Netherlands. Enviu, an environmental organization, partnered with Dutch architectural firm Doll to create this club in October 2008. This dance floor was a fusion of electronics, embedded software and smart durable materials.

Piezoelectricity is continuously being developed. Currently, researchers are working on using this technology for pedestrian and vehicular traffic, children's playgrounds, and amusement parks. Their goal is to harness the human movement and supply that power into the local grid.

Fabric that Can Harness Wind and Solar Energy

by Yerika Reyes

We have had the ability to produce fabrics that produce electricity from physical movement for a few years, but now researchers at Georgia Institute of Technology are developing a fabric that can gather solar energy and motion energy concurrently. The combination of these two generators of electricity into a textile will allow for developing clothes that can provide their own source of energy to power smartphones and global positioning systems (GPS). This fabric will alleviate the issue of charging devices while conducting research in the field because it can harness energy from the wind and sun.

This development was spearheaded by Zhong Lin Wang, a Regents professor in the Georgia Tech School of Materials Science and Engineering. This new material would be 320 micrometers thick woven together with strands of wool, and could be integrated into tents, curtains or wearable garments. To construct the fabric, Wang's team used a commercial textile machines to interlace together solar cells constructed from lightweight polymer fibers with fiber-based triboelectric nanogenerators. Triboelectric nanogenerators use a combination of the triboelectric effect and electrostatic induction to generate small amount of electrical power from mechanical motion such as rotation, sliding or vibration. Fiber-based triboelectric nanogenerators capture the energy created when certain materials become electrically charged after they come into moving contact with a different material. For the sunlight-harvesting part of the fabric, Wang's team used photoanodes made in a wire-shaped fashion that could be woven together with other fibers.

The fabric has proven that it can withstand repeated and rigorous use in early test, but needs to be tested for long-term durability. Additionally, there will further optimizing the fabric for industrial uses, including developing proper encapsulation to protect the electrical components from rain and moisture. This new fabric will prove to be invaluable both for those conducting research in a field and for those in need of charging their personal devices on the go.

https://qz.com/786461/scientists-have-invented-a-fabric-that-powers-mobile-devices-with-your-movements/
http://www.nature.com/articles/nenergy2016138

Veranu: Floors Generating Clean Energy

by Chloe Soltis

Veranu is a start-up based in Italy that was founded by Alessio Calcagni in 2012. The company started as a result of Calcagni's engineering thesis at the University of Cagliari called "Smart Energy Floor" (Veranu). Veranu develops and produces floor tiles containing piezoelectric materials that convert the kinetic power of walking steps into electricity (Veranu). Placing mechanical pressure on piezoelectric material creates a positive charge on the side that is compressed and a negative charge on the side that expands (Cleantechies). When the pressure is released, electricity moves across the material. Veranu has designed for the piezoelectric matter to be stored in a polylactic acid (PLA) box, which is made from renewable resources such as corn starch. Veranu covers the surface of the PLA box with common floor materials such as linoleum or ceramic tile so that the technology is invisible. Veranu's tiles are also waterproof and easy to install and maintain.

The electricity generated from the floors can be used for almost anything. Veranu has envisioned their floors being installed in large public spaces such as Rockefeller Center in New York City, where the floors would generate enough electricity to not only light-up the center but also power its iconic Christmas Tree and decorations during the holiday season.

Veranu is currently testing its latest tile prototypes in a space that has a large population of people pass through it every day (Veranu). The demo is testing the performance of the tiles and is introducing the public to a new form of alternative energy. While Veranu is excited to eventually sell their flooring, the company also plans to sell the data collected by the tiles. The floors will track how many people pass through their location over the course of a day and will be able to tell when the space is the busiest (Veranu). The company believes this type of data will be especially valuable to marketing companies. Veranu hopes their product will be sold publicly in the near future.

Veranu. http://www.veranu.eu.
"Piezoelectric Flooring: Harvesting Energy Using Footsteps." Cleantechies. July 8, 2015:
    http://cleantechies.com/2015/07/08/piezoelectric-flooring-harvesting-energy-
    using-footsteps/

Energy Storage

Tesla's Newest Project Involves Rooftop Solar Tiles with a Complimentary Powerwall 2 Upgrade

by Alejandra Chávez

In late March of 2017, the CEO and founder of Tesla, Elon Musk, revealed the company's newest project: rooftop solar tiles known as "Solar Roof." The project is in collaboration with SolarCity, the largest solar energy provider in the United States, and offers a complementary upgrade to Tesla's Powerwall home battery (a $5,000 value). The unveiling, which took place at Universal Studios on the set of Desperate Housewives, showcased an "inconspicuous aesthetic" with the Slate Glass Tile, Tuscan Glass Tile, Textured Glass Tile, and Smooth Glass Tile as the four different glass panels to choose from.

Though the solar tiles look like normal roofing tiles from the ground, they have a sophisticated three-layered anatomy: a "highly efficient" solar cell, a color louver film that gives the cells the ability to blend into the roof, and tempered glass that makes the solar cell tiles durable. To Musk, the tiles are 98 percent as efficient as traditional rooftop solar panels and have a "quasi-infinitive lifetime." Musk also claims that the cost of the tiles will still be cheaper than the cost of a normal roof plus additional solar panels, especially because these new solar roof tiles are designed to work with the upgraded Tesla home battery—The Powerwall 2. The Powerwall 2 is warranted for unlimited power cycles for up to 10 years, can store 14 kWh of energy, and offers a 5-kW continuous power draw (7-kW at its peak), which is enough to power an average household and an electric car The collaboration between the roof solar tiles and home battery are part of Tesla's larger mission to offer generation, storage, and transportation solar-powered solutions.

Though the company is set to start taking orders for its rooftop solar tiles in April, they have yet to give details on pricing, availability, and the installation of the Solar Roof.

[http://www.theverge.com/2016/10/28/13463236/tesla-solar-roof-battery-new-elon-musk].
[https://www.tesla.com/solarroof]. According

Cheah, Selina. "Tesla's new rooftop solar panels don't look like solar panels." Curbed. Vox Media, Inc., 28 Mar. 2017. Web. 9 Oct. 2017.

Curbed (http://www.curbed.com/2016/10/31/13478124/tesla-solar-roof-tiles-powerwall-batteries)

Golson, Jordan. "Tesla unveils residential 'solar roof' with updated battery storage system." The Verge. Vox Media Inc., 28 Oct. 2016. Web. 9 Oct. 2017.

The Verge (http://www.theverge.com/2016/10/28/13463236/tesla-solar-roof-battery-new-elon-musk)

"Tesla and SolarCity Announce Solar Roof." Solar Roof. Tesla, 2017. Web. 9 April. 2017.

Skyrmions: Less could be More

by Dominique Curtis

How would you like more storage on your electronics? Scientists at the University of Singapore have created an ultra-thin multilayer film with the potential to store large amounts of information. This nano sized film was created in collaboration between the researchers from Stony Brook University, Louisiana State University, and Brookhaven National Laboratory. The researchers state that "this is a critical step towards the design of data storage devices that use less power and work faster than existing memory technologies" (Pollard, 2017).

The ultra-thin multilayer film has properties of tiny magnetic whirls and they are called skyrmions. The skyrmions are the information carriers and are used for storing and processing data on magnetic media (Science Daily, 2017). Typically skyrmions only have spatial extent for a tens to a few hundred nanometres. Now with the idea of thinner film and multilayers there's potential to store much more information. Although researchers admit that due to it's tiny size it is difficult to image the nano-sized materials. In order to combat this limitation "researchers worked towards creating stable magnetic skyrmions at room temperature without the need for biasing magnetic field" (Science Daily, 2017).

The National University of Singapore researches found that they could have multilayered film composed of cobalt and palladium to stabilize skyrmion spin textures. They discovered that they could obtain clear contrast with sizes below 100 nanometres. To continue their developments the researchers are now looking into how their nanoscale skyrmions interact with each other and with electrical currents.

National University of Singapore. "Ultra-thin multilayer film for next-generation data storage and processing." ScienceDaily. ScienceDaily, 10 April 2017. <www.sciencedaily.com/releases/2017/04/170410085144.htm>.

Shawn D. Pollard, Joseph A. Garlow, Jiawei Yu, Zhen Wang, Yimei Zhu, Hyunsoo Yang. Observation of stable Néel skyrmions in cobalt/palladium multilayers with Lorentz transmission electron microscopy. Nature Communications, 2017; 8: 14761 DOI: 10.1038/ncomms14761

Coating Hot Batteries

by Dominique Curtis

Lithium ion batteries are important component of some of the things we use everyday, such as our laptops, cellphones, and even electric cars. Currently the negative charged side of lithium ion batteries are typically made up of graphite or other carbon-based materials. The carbon-based materials limits the performance ability because of the weight and energy density that could be used for more space.

Recently researchers have looked into using lithium metal. Lithium metal technologies have the ability to increase capacity of 5 to 10 times that amount of lithium ion technologies. This equates to 5 to 10 times more range for electric cars and 5 to 10 times the battery life for cell phones and laptops (Science Daily, 2017). Lithium Metal is also lighter and less expensive. Sounds great except one little flaw, Lithium Metal uncontrollably grows dendrites. According to researchers "The dendrites degrade the performance of the battery and also present a safety issue because they can short circuit the battery and in some cases catch fire" (Wu, 2017).

Researchers at the University of California Riverside have made significant advancements in solving this dendrites problem. The researchers found that by coating the battery with an organic compound called methyl viologen they can eliminate the dendrite problem and make the battery life last up to 3 times the amount of time it has in the past. The researchers state that the coating mechanism using methyl viologen is low cost and able to save battery life and it's compatible with the lithium-ion industry. This may be the start of more and better battery life and space. Although one researcher noted that this coating system does not prevent batteries from catching fire.

Haiping Wu, Yue Cao, Linxiao Geng, Chao Wang. In Situ Formation of Stable Interfacial Coating for High Performance Lithium Metal Anodes. Chemistry of Materials, 2017; DOI: 10.1021/acs.chemmater.6b05475

University of California - Riverside. "New battery coating could improve smart phones and electric vehicles." ScienceDaily. ScienceDaily, 17 April 2017. <www.sciencedaily.com/releases/2017/04/170417144938.htm>.

New Nanofiber May Provide Key to More Efficient Batteries

by Ethan Fukuto

In February 2017, materials science researchers at the Georgia Institute of Technology created a new nanofiber that may create stronger and more efficient rechargeable batteries. These double perovskite nanofibers can be used as catalysts for ultrafast oxygen evolution reactions, which are utilized in metal-air batteries as well as hydrogen-based energy sources. As the researchers note, electric cars in the future may use metal-air batteries, which can store more energy while being more compact than lithium-ion batteries currently in use, but lack a low-cost catalyst. While standard batteries have an anode and a cathode on either side and an electrolyte in-between, metal-air batteries use air as a cathode.

The nanofiber at Georgia Tech use composition tuning (or co-doping) and a unique crystal structure which improve catalytic activity by about 4.7 times. The material is about 20 nanometers in diameter, which is currently the thinnest diameter for any electrospun perovskite oxide nanofiber. During tests, the nanofiber's catalytic activity was around 72 times greater than a powder catalyst and 2.5 times greater than iridium oxide. Researchers see possibilities in the nanofiber aiding in the development and improvement of renewable energy systems, with the ability to store excess energy in, say, a wind or solar power system, which can produce hydrogen. While, as of now, storing that energy is neither inexpensive or efficient, the Georgia Tech nanofiber's catalytic activity can speed up the process in both water splitting and metal-air batteries. Tesla obtained a patent for charging metal-air battery technology in February as well, however issues regarding cost and the lifespan of such batteries have created some reservations in fully adopting metal-air batteries. The use of the Georgia Tech nanofiber, however, may mitigate cost issues while increasing efficiency. Its potential uses in larger projects brings the prospect of an efficient and inexpensive future for renewable energy.

Fehrenbacher, Katie. "A Battery Made From Metal and Air Is Electrifying the Developing World". Fortune. 23 May 2016. <http://fortune.com/2016/05/23/a-battery-made-from-metal-and-air-is-electrifying-the-developing-world/>

Georgia Institute of Technology. "New nanofiber marks important step in next generation battery development." ScienceDaily. ScienceDaily, 10 March 2017. <www.sciencedaily.com/releases/2017/03/170310121735.htm>.

Lambert, Fred. "Tesla obtains patent for charging metal-air battery technology that could enable longer range". Electrek. 13 Feb 2017. < https://electrek.co/2017/02/13/tesla-patent-metal-air-battery/>

Molecular 'Leaf' Can Store Solar Energy Without the Need for Panels

by Ethan Fukuto

A research team at Indiana University have engineered a molecule able to convert carbon dioxide into carbon monoxide using sunlight and electricity without solar cells. This molecule is an energy efficient means of turning carbon dioxide in the atmosphere into a useable, carbon-neutral fuel. In converting from carbon dioxide to monoxide, no more carbon is released back into the atmosphere; instead, the solar power used to create it is re-released. As the researchers report, the molecule, a well-defined nanographene-rhenium complex, requires the least amount of energy in carbon monoxide-producing catalysts. It can selectively electrocatalyze carbon dioxide reduction to carbon monoxide at -0.48V vs. NHE, which they state is the least negative potential on record. As they note, electron delocalization over the nanographene and the metal ion greatly reduces the potential needed for chemical reduction. A two-part system, the molecule absorbs energy from sunlight with the nanographene, which then drives a flow of electrons to an atomic rhenium 'engine', which binds to carbon dioxide and converts it to carbon monoxide. The efficiency of the molecule relies on its use of nanographene, a form of graphite commonly used as lead in pencils. Its dark color absorbs large amounts of sunlight and does not require a photosynthesizer to photocatalyze the chemical transformation. The molecule is able to use sunlight in wavelengths up to 600 nanometers, which covers a large portion of the visible light spectrum, while other complexes often used light in the ultraviolet range.

The primary researcher, Liang-shi Li, has used nanographene in the past to create efficient solar cell systems, but soon discovered that the light-absorbing quality of the graphene itself could drive the chemical reaction. Further research by Dr. Li will focus on making the molecule last longer and survive in non-liquid form, and to replace the rhenium atom with manganese, which is cheaper and more common. These additions may provide the tools needed to create efficient, inexpensive and clean energy sources.

Indiana University. "Chemists create molecular 'leaf' that collects and stores solar power without solar panels." ScienceDaily. ScienceDaily, 8 March 2017. <www.sciencedaily.com/releases/2017/03/170308135342.htm>.

Xiaoxiao Qiao, Qiqi Li, Richard N. Schaugaard, Benjamin W. Noffke, Yijun Liu, Dongping Li, Lu Liu, Krishnan Raghavachari, and Liang-shi Li. "Well-Defined Nanographene–Rhenium Complex as an Efficient Electrocatalyst and Photocatalyst for Selective CO2 Reduction". Journal of the American Chemical Society 2017 139 (11), 3934-3937. DOI: 10.1021/jacs.6b12530

The Skinny on Graphene

by Siena Hacker

Graphene has featured prominently in the news lately, and many are hailing it as the new "super material." Graphene currently claims the title of world's strongest material and the monomolecular sheets at its core are an estimated one million times thinner than paper. Researchers at Glasgow University recently discovered a new use for the super material: solar power electronic skin for prosthetic hands. According to a BBC article, researchers already developed an electronic skin that required a battery. However, the newest version of the graphene skin allows about 98% of light to pass through the surface and reach the photovoltaics underneath. The skin is also becoming increasingly capable of making sensitive pressure measurements. Dr. Ravinder Dahiya, one of the Glasgow researchers, says that amputees using the skin "are able to feel the contact pressure and temperature" when touching an object. Though the skin only requires 20 nanowatts of power per square centimeter, the researchers are currently exploring the possibility of diverting unused energy into a battery under the skin. Dr. Dahiya is hopeful that he will be be able to power the prosthetic limb's motors with renewable energy. He also says that advances in the technology could allow robots to more accurately gauge what they interact with. Robots with greater sensitivity capabilities might also be less likely to hurt humans or make errors. Amputees would benefit from the increased sensitivity as reduced weight of the prosthetic, which is significantly lighter without a battery. According to an Engadget article, the lack of battery also makes the prosthetics much cheaper. Researchers estimate that they could cost as little as $350 to produce, a cheap price compared to traditional prosthetic hands that usually cost tens of thousands of dollars. After taking on graphene skin for prosthetics, Dr. Dahiya said he hopes to use graphene technology to power vital devices, like glucose monitors, for patients without access to electricity.

Engadget: https://www.engadget.com/2017/03/23/artificial-skin-with-solar-cells-could-power-prosthetics/

BBC: http://www.bbc.com/news/uk-scotland-39353751

Unlocking Graphene's Superconductivity
by Parker Head

Researchers at Cambridge University recently conducted experiments, the results of which suggest graphene's potential to be used as a superconductor on its own. Graphene is a layer of carbon so thin it is two-dimensional. So far the commercial application of graphene has already brought about increases in energy efficiency to lightbulbs, with graphene lightbulbs having longer life-spans than LEDs, and projected future innovations in motorcycle safety equipment and dental ware as well.The thinness of the material, coupled with its strength and conductive properties, make it well-suited for a multitude of applications, as seen in the variety of the uses mentioned above.

But, in order to use graphene for superconductive purposes it has required pairing it with other materials, doping it with or placing it on superconductive materials. This additional material compromises graphene's superior weight-strength ratio. That is, until this most recent research was able to unlock graphene's inherent superconductivity. The researchers at the University of Cambridge began by coupling graphene with the well-studied superconductor praseodymium cerium copper oxide (PCCO). The researchers' familiarity with PCCOs superconductive characteristics is important, because when graphene was coupled with it the observed superconductivity changed from d-wave to a completely different type, possibly the highly illusive p-wave. The exact type of superconductivity is unknown, and the p-wave type cannot be definitively identified because it is still being proven to exist, but the fact that the observed superconductivity was a type different than PCCOs well-known type means the source of superconductivity was graphene.

If graphene's superconductivity could be activated in isolation it would mean the existence of a superconducting material one-atom thick and stronger than steel. This would allow for new advances in molecular electronics and the creation of superconducting materials for quantum computing.

[https://phys.org/news/2017-01-graphene-superconductivity-awakens.html].

[http://newatlas.com/graphene-pwave-superconductor-cambridge/47500/].
[http://newatlas.com/light-bulb-graphene-first-commercial-consumer-application/36812/],
Jeffrey, Colin. 2017. Scientists unleash graphene's innate superconductivity. New Atlas.
        Jan 21, 2017. http://newatlas.com/graphene-pwave-superconductor-
        cambridge/47500/
Jeffrey, Colin. 2017. Light bulb set to be graphene's first commercial consumer application.
        New Atlas. March 31, 2015. http://newatlas.com/light-bulb-graphene-first-
        commercial-consumer-application/36812/
Graphene's sleeping superconductivity awakens. PhysOrg. Jan 21, 2017.
        https://phys.org/news/2017-01-graphene-superconductivity-awakens.html

Your Finger as Phone Charger

by Parker Head

Pennsylvania State University researchers are in the process of developing a novel new material for use in touch-screen electronics that could turn your finger into a charging source for your phone. Lynda Delacey reports that the research into this alternative phone battery technology is funded by Samsung, a company with obvious incentive to explore new methods of mobile electronic energy production. The material would convert the mechanical energy of the user's finger touching the screen into electrical energy that could power the phone. The research team's goal is for this material to be able to produce up to 40 percent of the device's energy.

The material itself is actually a layered-cake sort of composite that consists of nanocomposite electrodes of different charges separated by a polycarbonate membrane and sandwiched between two platinum foils. This type of energy harvesting is known as piezoelectric energy conversion, and, until this new material, would not have been efficient enough for use with the sort of mechanical force applied by a phone user's finger. This is because the movement of a finger is relatively slow compared to the type of forces usually implemented with piezoelectric energy conversation, movements of much higher frequencies, viz. greater than 10 vibrations per second.

The task of the Penn State researchers that developed this new material was to create a piezoelectric energy converting material that worked with slower movements, specifically the swipes and taps of a finger on a phone screen, as opposed to movements so rapid as to be described as vibrations. The ions in the electrodes of the material, when distorted by a force, diffuse across the membranous separator, which creates a current. But, a potential is created as well, which opposes this diffusion until equilibrium between the separator is restored. In total, this cycle operates at one-tenth Hertz, about once every 10 seconds.

[http://onlinelibrary.wiley.com/doi/10.1002/aenm.201601983/full].
[http://newatlas.com/finger-movements-could-power-negt-gen/47135/].
Delacey, Lynda. 2016. Slo-mo energy harvesting could see finger presses powering
        touchscreen devices. New Atlas. December 29, 2016.
        http://newatlas.com/finger-movements-could-power-negt-gen/47135/
Hou, Ying et al. 2016. Flexible Ionic Diodes for Low-Frequency Mechanical Energy
        Harvesting. Advanced Energy Materials. November 16, 2016.
        http://onlinelibrary.wiley.com/doi/10.1002/aenm.201601983/full
Lynda Delacey's Twitter: https://twitter.com/LJDelacey

Hydrogen Fuel at Hand

by Parker Head

Hydrogen as a widely used, industrial and commercial, clean fuel source has obvious drawbacks; hydrogen is hard to handle and transport safely, and it burns at a very high temperature. But research taking place at Waseda University in Japan has produced a new material that has been shown to be able to more safely and conveniently store hydrogen.

A new hydrogen-absorbing polymer has proven to be safe to carry and handle when filled with hydrogen. The polymer is made from the ketone fluorenone. The hydrogen is fixed within this polymer using water at room temperature. When this is heated to 176 °F with an aqueous iridium catalyst the polymer releases the hydrogen. The key features here are the fact that the hydrogen is reversibly bound to the fluorenone and is released at just 176 degrees, a relatively "mild" temperature, and with only a water-based catalyst rather than one made from precious metals or using harsh solvents.

The polymer can be molded as a plastic sheet. The plastic sheet containing the hydrogen is non-flammable and safely pliable, in stark contrast to the bulky and expensive pressurized tanks currently used. Soon hydrogen could be used domestically as well as industrially, with the plastic polymer being safe, lightweight, and easily transportable. These features directly address some of the main hindrances of hydrogen used as an energy source. Delacey postulates some potential uses for this newly mobilized version of hydrogen fuel as being: pocket-sized hydrogen releasing fuel cartridges, mobile hydrogen generators, and even on-site hydrogen supply sources for bases as remote as the Moon or Mars.

[http://www.nature.com/articles/ncomms13032].

Delacey, Lynda. 2016. Finally – a safe way to carry hydrogen fuel in your pocket. New Atlas. December 1, 2016. http://newatlas.com/new-polymer-stores-hydrogen-safely/46686/

Kato, Ryo et al. 2016. A ketone/alcohol polymer for cycle of electrolytic hydrogen-fixing with water and releasing under mild conditions. Nature Communications. September 30, 2016 http://www.nature.com/articles/ncomms13032

Collie, Scott. 2016. Honda Clarity makes the case for hydrogen with 366 mile range. New Atlas. October 25, 2016. http://newatlas.com/honda-clarity-fuel-cell-epa-range/46108/

Paper Battery Powered with Dirty Water

by Parker Head

Researchers at Binghamton University have revealed the newest iteration of their paper-based and bacteria-energized battery. Like past prototypes, the battery is designed with an origami inspired folding capability that produces different shapes, resulting in different power outputs. The newest model is the thinnest yet, mounted on just a single sheet of paper. The battery is composed of primarily paper and water, making it biodegradable. This, coupled with its portability, make it a promising alternative to less eco-friendly, low-power batteries that are widely used in personal biosensors, such as blood-sugar monitors.

The battery is powered by cellular respiration carried out by the bacteria living in the water. On one side of the piece of chromatography paper-thin reservoirs of the bacteria-infested water are contained within a conductive polymer. The other side has a wax coated strip of silver nitrate on it. When the paper is folded, and the two sides come into contact with one another, cellular respiration takes place and generates electricity. A 2 x 3 grid of chromatography paper outfitted with these components generates 31.51 microwatts at 125.53 microamps when folded, and a 6 x 6 grid was able to generate 44.85 microwatts at 105.89 microamps. While these numbers may seem insignificant, being not enough to charge a cellphone, they are plenty high enough to power biosensors such as blood sugar monitors.

The low production cost, portability, and biodegradability of the paper batteries make them worth continued research. While they may not power a phone or a car, the products they do power are potentially lifesaving. Professor Choi, one of the lead researchers of the project, says of the medical applications that, "Stand-alone and self-sustained, paper-based, point-of-care devices are essential to providing effective and life-saving treatments in resource-limited settings".

[http://newatlas.com/bacteria-powered-paper-battery/47073/].
Irving, Michael. 2016. Paperback bacteria biobattery folds for different power levels. New Atlas. December 21, 2016. http://newatlas.com/bacteria-powered-paper-battery/47073/
Gao, Yang and Choi, Seokheun. 2016. Stepping Toward Self-Powered Papertronics: Integrating Biobatteries into a Single Sheet of Paper. Advanced Energy Materials. October 18, 2016.
http://onlinelibrary.wiley.com/doi/10.1002/admt.201600194/full
Binghamton University's Twitter:
https://twitter.com/binghamtonu?ref_src=twsrc%5Egoogle%7Ctwcamp%5Eserp%7Ctwgr%5Eauthor

Rhubarb Battery

by Parker Head

Researchers a Harvard University's John A. Paulson School of Engineering have invented a new battery that is competitive with lithium-ion ones and is scalable to domestic and industrial uses. But, the real novelty is that it runs on a compound similar in chemical structure to molecules that occur in rhubarb.

This new battery generates an electric current through the process of ion exchange that occurs between two specialized liquids as they are pumped from tanks and flow adjacent to one another across a membrane. While this battery type, known as a flow battery, is not new, the efficiency and sustainability of the materials used by the one recently created by the Harvard research team is. Previously, the liquids in flow batteries were hazardous, making them dangerous, expensive, and environmentally unsound. The new battery, on the other hand, uses a pH-neutral water solution as its liquid base.

The breakthrough of this research is in the particular chemicals dissolved in the water base. Until recently, flow batteries relied on the ion exchange between metal-based solutions; however, the new battery replaces hazardous metal-based solutions with a water-quinone solution. Quinones are cheap, abundant, carbon-based molecules that are similar to molecules that store energy in plants. The research team tested more than 10,000 different quinone molecules in order to determine which would be the best for a battery application. The quinone finally used in the battery resembles quinones found in rhubarb.

Flow batteries are ideal for energy storage from wind and solar power. The energy is stored in chemical form in tanks of solution which can be charged or discharged by pumping the solution over a membrane when the need arises. Their storage capacity is limited only by the amount of solution available, unlike solid-electrode batteries which, "can maintain peak discharge power for less than an hour before being drained". The new quinone battery losses only one percent capacity every 1,000 cycles. This breakthrough efficiency, paired with its lessened cost and organic base, make it a revolutionary discovery in terms of how renewable energy can be stored and utilized in an equally environmentally conscious way.

[https://cleantechnica.com/2017/02/13/harvard-u-trump-mega-rhubarb-energy-storage-device-beats-coal/].

[https://www.seas.harvard.edu/news/2014/01/organic-mega-flow-battery-promises-breakthrough-for-renewable-energy].

Casey, Tina. 2017. Harvard U. to Trump: Our Mega Rhubard Energy Storage Device Beats Your Coal. Clean Technica. February 13, 2017. https://cleantechnica.com/2017/02/13/harvard-u-trump-mega-rhubarb-energy-storage-device-beats-coal/

N.a. 2017. Organic mega flow battery promises breakthrough for renewable energy. Harvard. January 8, 2014. https://www.seas.harvard.edu/news/2014/01/organic-mega-flow-battery-promises-breakthrough-for-renewable-energy

Artificial Sun

by Cybele Kappos

German scientists from the German Space Center (DLR) have created the largest artificial sun in the world and have just turned it on. The aim of their experiment is to create climate-friendly fuel. The 'sun' consists of 149 film projector spotlights that have been modified to increase power and efficiency and in total cost 3.5 million euros. The device can produce around 10,000 times the intensity of natural sunlight on Earth's surface. When the light from the spotlights is concentrated in one spot, which measures 20 x 20 cm, they can produce temperatures of about 3,500°C. The experiment is housed in a protective radiation chamber because of the severe danger it poses to anyone who might enter the room.

By concentrating the light, the scientists are striving to power a reaction that would produce hydrogen fuel. There are already solar stations that concentrate sunlight onto water to produce steam that in turn spins turbines in order to generate electricity. This experiment, titled Synlight, is exploring the possibility of creating a similar reaction so that they can extract hydrogen from water vapor. The hydrogen can be used as a fuel source for airplanes and cars. Synlight currently consumes an enormous amount of energy. Four hours of usage consumes the same energy that a four-person household does in an entire year. Experimenting with natural sunlight poses difficulties such as the limited hours of daylight as well as cloudy days and shorter days in the winter, which hinder the capacity to experiment. Synlight is a temporary substitute. In the future, the scientists hope that natural sunlight could be controlled in order to produce hydrogen in a carbon-neutral way.

They are attempting to create a CO_2-free fuel that would be sustainable since climate change is a major concern. Additionally, they are studying how different materials age under exposure to UV rays.

Devlin, Hannah. Let There be Light: Germans switch on 'largest artificial sun'. 23 March 2017
https://www.theguardian.com/science/2017/mar/23/worlds-largest-artificial-sun-german-scientists-activate-synlight
http://newatlas.com/dlr-artificial-sun/48579/

Green Climate Fund supports Renewable Energy Development and Storage Program in the Pacific Islands

by Genevieve Kules

In late December of 2016 the Green Climate Fund (GCF) granted $17 million to the Asian Development Bank's (ADB) renewable energy program to take seven countries in the Pacific Islands from a current overall 15% use of renewable energy to 50% in the next fifteen years. The analysis this project is based on comes from an initial report by the International Renewable Energy Association in 2013 outlining viable alternatives to fossil fuel combustion engines which are currently the primary source of energy production in the Pacific Islands.

The primary forms of energy generation will be solar, hydropower, and wind, as well as a series of mini-grid projects and energy storage facilities. While each country faces its own specific challenges in this transformation, this grant encourages the countries involved to learn from each other. It also boosts the involvement of the private sector in renewable energy production.

The Cook Islands' capital Rarotonga is currently only at 15% renewable energy production, but by the year 2020 it is expected to be at 100%. The Cook Islands have already seen an increase in private sector and individual home investments in renewable energy power sources. If the Cook Islands are an indication of the trajectory of all seven nations being funded by this grant, it could be an important step for not only the Pacific Islands but also for the atmosphere.

The GCF and ADB anticipate this grant to save between 95.6 thousand and 120 thousand tons of carbon dioxide from entering the atmosphere annually. The GCF has stated that this project will particularly affect women. They will have better access to electricity in their homes, even in rural places due to mini-grid systems. It aims to meet the UN sustainability goal of affordable and clean energy. In addition to the Cook Islands, other countries involved in this project are Papau New Guinea, Samoa, Tonga, Republic of Marshall Islands, Federated States of Micronesia, and Nauru. Though still in need of funding (another $400 million) the project appears to be starting off strong.

Ngabung, Kiwiana. "Financial Boost to Allow for More Integration of Renewable Energy in Pacific Region." EMTV Online. EMTV, 10 Jan. 2017. Web. 6 Feb. 2017. <http://www.emtv.com.pg/news/2017/01/financial-boost-to-allow-for-more-integration-of-renewable-energy-in-pacific-region/>.

"Pacific Islands Renewable Energy Investment Program." Green Climate Fund. N.p., 23 Dec. 2016. Web. 6 Feb. 2017. <http://www.greenclimate.fund/-/pacific-islands-renewable-energy-investment-program>.

Shute-Trembath, Sally, and Ayun Sundari. "GCF Supports ADB to Accelerate Shift to Renewable Energy in the Pacific." Asian Development Bank. N.p., 19 Dec. 2016. Web. 6 Feb. 2017. <https://www.adb.org/news/gcf-supports-adb-accelerate-shift-renewable-energy-pacific>.

Shumkov, Ivan. "GCF to Support Renewables on Pacific Islands." Renewables Now. N.p., 20 Dec. 2016. Web. 6 Feb. 2017. <https://renewablesnow.com/news/gcf-to-support-renewables-on-pacific-islands-551376/>.

A Single Substance can Convert Heat, Movement and Light into Electricity

by Genevieve Kules

The future brings many innovations, one of which is a material that can convert at least three types of natural resources and occurrences into a an electric force. KBNNO is a single material that can use sunlight, movement, and heat to produce electricity, potentially enough to run your small personal devices like watches and cell-phones.

How does it work? Until now researchers have only been able to harvest energy from one source at a time, without the devices becoming bulky and impractical for daily use. KBNNO, on the other hand, has the properties of a ferroelectric, pyroelectric, and piezoelectric substance as well as is a photovoltaic perovskite solution. Ferroelectricity has a permanently polarized field that gains electricity when its electrons move around due to a variety of factors. A pyroelectric material generates electricity when it heats, and a piezoelectric material converts movement into electricity. The final one mentioned is a photovoltaic property which means it can create energy from the sun.

The possibility of having one material that can do all three of these types of electric conversion is a promising discovery. One of the problems I often have on a day to day basis is running out of battery on my cell-phone. I am forced to carry around an extra charged battery and my chord in order to get through the day. If I had a device made out of this substance it would constantly be charging itself—or I would be charging it with my movements. KBNNO is still new, but it is a breakthrough in electricity-generating technology. Although right now the authors are only discussing its power to charge small devices, I wonder what new breakthroughs similar to this will bring for cars and larger electric vehicles and technologies in the future.

American Institute of Physics. "Material can turn sunlight, heat and movement into electricity -- all at once: Extracting energy from multiple sources could help power wearable technology." ScienceDaily. ScienceDaily, 7 February 2017. <www.sciencedaily.com/releases/2017/02/170207142711.htm>.

Bai, Yang, Tuomo Siponkoski, Jani Peräntie, Heli Jantunen, and Jari Juuti. "Ferroelectric, Pyroelectric, and Piezoelectric Properties of a Photovoltaic Perovskite Oxide." Applied Physics Letters. AIP Publishing, Feb. 2017. Web. 14 Feb. 2017. <http://aip.scitation.org/doi/full/10.1063/1.4974735>.

Sodium-Ion Battery has Massive Energy Storage Potential

by Genevieve Kules

Lithium-ion batteries are probably in your cell phone, laptop, electric car, or any device that is rechargeable. This could be about to change. New research on sodium-ion batteries combined with an anode material, such as borophene, has created longer life cycles for the battery. While sodium-ion batteries may not replace lithium-ion batteries altogether, they do show promise for long term renewable energy storage.

One sodium-ion battery on the market is called the Aquion Energy M110-LS83 Battery sold by Wholesale Solar. There are different versions of this battery: one for standard residential use, and this larger one for off-grid home and commercial energy storage. The Aquion Energy Battery is modular which allows for differing storage sizes and shipping abilities. The battery is big in physical size and price: 45 by 52 by 40 inches and 3,309 lbs at $17,540 - definitely not portable or affordable...yet.

Planet Earth has a surplus of sodium, whereas lithium is only found in parts of South America. Exploiting the earth of lithium from one certain place is not sustainable, but desalination processes could leave us with an abundance of sodium that could be easily used to make sodium-ion batteries.

Lithium-ion batteries are not only bad for the environment because of extraction, but because of disposal as well. They can overheat and explode because of power distribution malfunctions, and are difficult to dispose of due to the harmful chemicals that make the battery operational. Sodium-ion batteries are better for the environment than their alternatives because they do not contain heavy metals or toxic chemicals. Additionally, they are far safer than lithium-ion batteries because they are not flammable, explosive, or corrosive.

Though they are not yet at the capacity of lithium-ion batteries, they are well on their way to replacing them for large energy storage purposes.

American Chemical Society. "Making sodium-ion batteries that last." ScienceDaily. ScienceDaily, 15 February 2017. <www.sciencedaily.com/releases/2017/02/170215101441.htm>.

Hellemans, Alexander. "Here's a Peek at the First Sodium-ion Rechargeable Battery." IEEE Spectrum: Technology, Engineering, and Science News. N.p., 03 Dec. 2015. Web. 17 Feb. 2017. <http://spectrum.ieee.org/energywise/energy/renewables/a-first-prototype-of-a-sodiumion-rechargeable-battery>.

"Aquion Energy M110-LS83 Sodium-Ion Battery 580Ah 48V USA Battery."WholesaleSolar.com. N.p., n.d. Web. 17 Feb. 2017. <https://www.wholesalesolar.com/9949505/aquion-energy/batteries/aquion-energy-m110-ls83-sodium-ion-battery-580ah-48v-usa-battery>.

"Borophene Could Be an Extraordinary Sodium Anode Material for Sodium-based Batteries."Phys.org. Science X Network, 24 June 2016. Web. 20 Feb. 2017. <https://phys.org/news/2016-06-borophene-extraordinary-sodium-anode-material.html>.

Fractals Used for Solar Energy Storage in New Graphene-based Electrode

by Genevieve Kules

For the first time, total reliance on solar does not sound so far away. Scientists from RMIT Melbourne, Australia found that fractals within the western Swordfern are an effective model for storing solar energy within supercapacitors. They created a graphene-based electrode that could raise the storage capacity of supercapacitors by up to 30%. Supercapacitors allow energy to be emitted faster than conventional lithium-ion batteries, but storage capacity has always been their problem. This graphene-based electrode could be that solution.

If scientists are able to scale this electrode to a reasonable size, it could no longer be necessary to charge our phones and portable electronics in the wall. Lead scientist, Dr. Litty Thekkekara, explains that the electrode is meant to be used with flexible thin-film solar panels, which gives it a multitude of possibilities for use. Paired with thin-film solar cells, electronic devices would be able to charge whenever exposed to the sun. The next step is to develop flexible solar film that can be applied to the outsides of phones and various technologies.

One technology that contributed to this research is the transparent electrode, or more specifically the graphene-based transparent organic light emitting diode developed by a Korean research team led by Professors Seunghyup and Tae-Woo Lee. The transparent nature of this electrode will allow it to collect energy without disrupting light passing through surfaces such as phone screens and windows. Additionally, OLEDs have a future for foldable screen technologies. They could be placed on skin and used for health monitoring or potentially rollable smartphones and tablets.

RMIT University. "Bio-inspired energy storage: A new light for solar power: Graphene-based electrode prototype, inspired by fern leaves, could be the answer to solar energy storage challenge." ScienceDaily. ScienceDaily, 31 March 2017. <www.sciencedaily.com/releases/2017/03/170331120317.htm>.
"Graphene-based Transparent Electrodes for Highly Efficient Flexible OLEDS." Phys.org. Science X Network, 3 June 2016. Web. 4 Apr. 2017. <https://phys.org/news/2016-06-graphene-based-transparent-electrodes-highly-efficient.html>.

Storing Solar Energy Increases Overall Energy Consumption Because of Inefficiencies

by Nina Lee

On January 30, 2017, the University of Texas at Austin released a study suggesting that at-home storage of solar energy may not actually be the most efficient way to power your house. According to UT's Cockrell School of Engineering, among homes that have solar panels installed, those who stored energy for nighttime usage consumed more energy than those who redistributed their energy to utility grids. While energy storage is commonly understood to be inherently clean, when looked into in more detail, the study shows that about fifteen percent of the energy meant to go into home battery systems are actually lost due to inefficiencies. Because of this, the total amount of solar energy produced by the house ends up declining over that of a similar solar array on a house without a battery.

The study looked at 99 Texas households during the year 2014, all part of a smart grid test bed managed by UT Austin's renewable energy and smart technology company, Pecan Street Inc. Researchers estimated that homes with energy storage for their solar panels increase their yearly energy consumption by 324 to 591 kilowatt-hours. In addition, researchers have also found that with regards to Texas' current fossil-fuel powered grid-mix, energy storage indirectly increases emissions of carbon dioxide, sulfur dioxide, and nitrogen dioxide.

Michael Webber and Robert Fares, Cockrell School alumnus and co-authors of the study, have made sure to point out that storing energy from solar panels is still much more efficient than having no solar panels at all. In terms of overall energy usage, solar energy storage reduces grid demand by 8 to 32 percent and the magnitude of solar power injections to the grid by 5 to 42 percent, cutting the greater need for energy production.

https://www.sciencedaily.com/releases/2017/01/170130133322.htm
https://www.sciencedaily.com/releases/2017/01/170130133322.htm
https://blogs.scientificamerican.com/plugged-in/storing-solar-energy-in-the-home-can-increase-energy-consumption-emissions/

Freezing Energy in Ice Bear Batteries

by Nina Lee

Harnessing the power of nature's coolant, IceEnergy has created a series of smart ice batteries called Ice Bears. On February 14, 2017, IceEnergy announced the release of another line of Ice Bears called the Polar Bear. While the previous Ice Bear products have been focused on residential and smaller commercial buildings, the Polar Bear will cater to larger businesses facilities that require large amounts energy for of refrigeration. All Ice Bear batteries can be added to existing commercial air conditioning units and either air duct systems or mini-split air conditioner systems, requiring only the installation of an ice coil to the machine.

The Ice Bear works similarly to a regular battery, but it stores electrical energy in the form of ice. The battery "charges" overnight, making ice during off-peak hours to reduce electricity costs. The Ice Bear then takes over during the hottest part of the day, which is also when energy is the most expensive. Cold refrigerant from the Ice Bear is transferred to the existing air conditioning unit, allowing the warm air from the building to cool as it passes through the ice coils. The refrigerant, now warm, is transferred back to the Ice Bear. The Ice Bear in effect replaces the job of the compressor in the air conditioning unit. This system can continue for about six hours before the regular air conditioning unit needs to take over again.

The Ice Bear battery reduces electricity usage during peak hours by 95% and costs less than half the price of lithium ion batteries of the same capacity. Customers on the average save 40% on their cooling bills, and the Ice Bear can last at least twenty years with no degradation. In terms of its environmental benefits, Ice Bear storage reduces CO_2 emissions by up to 40% and NOx emissions by up to 56%.

https://www.ice-energy.com/
http://www.marketwired.com/press-release/ice-energy-extends-award-winning-multi-patented-ice-battery-technology-commercial-refrigeration-2195752.htm

LuminAID's New PackLite Max is a 2-in-1 Lantern and Charging Device

by Nina Lee

After the devastating 2010 earthquake in Haiti, architecture graduate students Andrea Sreshta and Anna Stork designed a product to aid the relief efforts. The two focused on the need for light in disaster-stricken areas, creating a design for an ultra-light, inflatable solar-powered lantern. The lantern's inflatable quality allows for the products to be shipped out in mass quantities. The products were such a success that the Sresha and Stork were featured on the television show *Shark Tank*.

Since the rise of LuminAID, the products have expanded from emergency usage to outdoor recreational uses as well. However, remaining true to their roots, LuminAID's products are on sale and are also a part of a program called "Give Light, Get Light." This program partners with global non-profit organizations to distribute one LuminAID lantern for every lantern purchased through the program.

The newest product by LuminAID is an expansion on the original design called the PackLite Max Phone Charger. The organization put out a Kickstarter campaign that ended in early March, significantly surpassing their goal. The PackLite max is also a lantern, capable of producing light for up to 150 hours from solar energy, but also has an additional 2,000 mAh lithium-ion battery capable of charging small electronic devices. The PackLite Max weighs 8.5 ounces and is only one inch thick when deflated, but can expand into a six inch cube when inflated.

The PackLite's new feature is especially useful for refugees and displaced people, as having access to communication is almost as important as having access to sustainable light sources. LuminAID has already distributed their products to refugees in Greece and other locations in Jordan, and they plan on partnering with the organization SCM Medical Missions to send more PackLites to Syrian refugees in Jordan at the end of March.

http://www.digitaltrends.com/outdoors/luminaid-phone-charger-lantern/
https://www.treehugger.com/gadgets/35-inflatable-waterproof-solar-lantern-will-also-charge-your-phone.html
http://www.chicagotribune.com/bluesky/originals/ct-shark-tank-luminaid-kickstarter-bsi-20170207-story.html
https://luminaid.com/

Stanford Lab Creates Low Cost Aluminum Battery

by Kieran McVeigh

In the first week of February 2017 a Stanford Chemistry lab announced they had created a new low cost underlined aluminum battery, which they believe could be a cheap and safe way to store excess solar energy created during the daylight hours for use at night. The new battery, designed by Stanford Professor Hongjie Dai and PhD student Michael Angell is made from graphite, aluminum, and urea, which Professor Dai describes as, three of the "cheapest and most abundant materials" on earth.

One thing that makes this new battery unorthodox is its use of urea. Urea is a chemical compound found in fertilizer and animal urine, not the stuff that comes to mind when you think of a battery. By using, urea researchers were able to improve the design of a previous battery they created using similar materials in 2015. The previous battery however used "an expensive electrolyte", which urea was substituted for in current design. The new battery, by researchers estimates is 100 times cheaper than the previous generation, which they suggest upon commercialization will make this new battery much more affordable than current lithium-ion batteries.

Initial tests suggest that, the batterie's quality is not compromised by its small price tag. The battery scored a 99.7% on a measure of efficiency called the columbic efficiency rating, which measures how much charge escapes the battery while charging. The article suggests that this efficiency is comparable to lithium ion batteries, and stresses that since the materials in this battery are non-flammable this battery is also safer then current lithium-ion batteries. Although the creation of this battery has just been announced, AB systems, a company founded by Professor Dai, is already working on a commercial version of it. As we look forward to seeing this battery hit the market we can only hope it lives up to the hype the researchers are building around it.

Markham, D. (2017, February 10). New aluminum battery with urea electrolyte could be a low-cost renewable energy storage solution. Retrieved February 13, 2017, from http://www.treehugger.com/clean-technology/new-battery-made-aluminum-graphite-and-urea-could-be-low-cost-home-solar-storage-solution.html

Flynn, J. (2017, February 08). Stanford engineers create a low-cost battery for storing renewable energy. Retrieved February 13, 2017, from http://news.stanford.edu/2017/02/07/stanford-engineers-create-low-cost-battery-storing-renewable-energy/

A Battery Made Out Of Ice?

by Kieran McVeigh

In early February of 2017 the Southern California Public Power Authority bought one megawatt of Ice batteries. Ice batteries are a "battery" system designed to lower air conditioning costs of residential and commercial buildings, by making and using ice to cool buildings for part of the day, as opposed to normal air conditioning units. Companies have been making ice batteries for over ten years, but the market for ice batteries has recently seen increased interest including many recent investments from venture capital firms.

These ice batteries work by making a big block of ice, using power from the grid, and then using the cold air that comes off of that ice to cool buildings. This method of cooling does not sound particularly efficient, the metaphorical equivalent of cooling your house down by keeping the freezer door open, but where the ice battery has its advantage is that after the ice is created; a building can be cooled with the ice battery for up to 6 hours while using five percent of the energy that traditional air conditioning units use. This efficiency allows building and homeowners to reduce their energy usage during peak hours, and then charge the ice battery during non-peak hours. By some estimates this can reduce cooling costs by as much as 40%.

Although ice batteries seem a positive way to alleviate energy usage at peak hours, it seems they would contribute to the problem of needing to store incoming solar energy for when we need it. As the manufactures suggest charging the ice battery night, when solar power tends not be coming in, if there were some way to limit this inefficiency then I could see these ice batteries being a more viable solution. It seems the ice battery is relevant to current problems with energy production and demand but as we (hopefully) move closer to renewable sources of energy the ice batteries appeal may melt away.

Fehrenbacher, Katie. "Tesla's CTO Just Backed A Startup That Makes An Ice Battery." Tesla CTO Just Backed A Startup That Makes An Ice Battery | Fortune.com. Fortune, 04 Aug. 2016. Web. 20 Feb. 2017. <http://fortune.com/2016/08/03/tesla-cto-ice-battery/>.

McDonald, Michael. "Are Ice Batteries The Future Of Energy Storage?" OilPrice.com. OilPrice.com, 16 Feb. 2017. Web. 20 Feb. 2017. <http://oilprice.com/Energy/Energy-General/Are-Ice-Batteries-The-Future-Of-Energy-Storage.html>.

China Develops First Hydrogen-Powered Tram
by Byron R. Núñez

China is notorious for being the number one carbon polluter in the world. The country's carbon footprint and emissions grew almost exponentially during the earlier part of the 21st century and are now slowing down as economic growth stagnates. Many private Chinese companies and government agencies are hoping to reduce harmful and plentiful greenhouse gas emissions by tapping into renewable sources of energy. The Foshan government located in northwest China, for example, contracted Sifang to provide hydrogen-powered trams.

In March 2015, Sifang produced seven hydrogen fuel cell hybrid trams that entered the passenger service line in Qingdao, China. However, after two years of research and development, the tram in Foshan will be powered entirely by hydrogen fuel cells. These trams will have a carrying capacity of 380 passengers, reach speeds of 70 kilometers per hour, have a 100-kilometer range, and only take 3 minutes to refuel. The trams will not produce nitrogen oxides because the temperature of the fuel cells will be kept under 100°C. Sifang will supply 8 of these hydrogen fuel cell trams to a route currently being developed in Foshan. The total cost of the project is reported to be around $110 million.

Ground broke on the project on February 27, 2017 and by 2018 it is expected that these trams will be operating on a 7-kilometer network that will consist of 10 stations and a refueling stop. The hydrogen fuel cells for the trams will be developed by a Canadian company called Ballard Power Systems under a $6 million agreement with Sifang and the Foshan government. Sifang hopes to make its trams widely accessible in cities thought the country to help alleviate the harmful impact that coal has had on the air quality and environment of China.

The National Post
http://news.nationalpost.com/news/worlds-biggest-polluters-china-u-s-and-india-help-drive-carbon-dioxide-levels-to-record-high
FuelCellsWorks
https://fuelcellsworks.com/news/chinafoshan-hydrogen-fuel-cell-tram-contract-signed

Axiom Exergy: Refrigeration Battery is an Energy Storage Solution for Supermarkets

by Chloe Soltis

Refrigeration is a major expense for supermarkets. It is estimated that supermarkets consume three times the amount of electricity of other retailers and spend 60% of their energy bills on electricity for refrigeration (Axiom Energy). In addition, electricity rates can raise by 400% during the peak hours of the day 12 – 6 PM when refrigeration is needed most. Axiom Exergy recognized this problem and developed the Refrigeration Battery.

The Refrigeration Battery is an intelligent energy storage solution that stores energy or "cold" when prices are low and then uses the stored energy during the peak hours of the day. The Refrigeration Battery consists of three pieces that can be installed into any central refrigeration unit through retrofitting, which means the original cooling system does not need to be reprogrammed or reconfigured. The energy is stored in an insulated thermal storage tank filled with saltwater that freezes in order to store the cold. The tank can be installed underground or in the parking lot of a store. The electronic dashboard and system integrator are the "brains" of the Refrigeration Battery and monitor when the system should use cold stored energy in response to surging prices, weather patterns, or grid programs such as demand response (Axiom Exergy and Utility Dive). This system decreases electricity spending up to 40% during peak hours.

Axiom Exergy has recently partnered with Wal-Mart to install a test version of the Refrigeration Battery at a San Diego location. Walmart has claimed that it eventually wants to be a business using 100% renewable energy, and the company believes Axiom Energy's Refrigeration Batteries could be part of this solution (Utility Dive). Supermarkets are in a competitive market with thin margins, which makes products like the Refrigeration Battery an attractive investment for saving money.

Axiom Exergy. http://www.axiomexergy.com/index.html.

Maloney, Peter. "Walmart, Axiom Exergy partner for refrigeration energy storage project." Utility Dive. April 17, 2017: http://www.utilitydive.com/news/walmart-axiom-exergy-partner-for-refrigeration-energy-storage-project/430487/.

Long-Lasting Battery Solution

by Justin Wenig

Researchers at Harvard University have developed a non-toxic, recyclable battery that stores energy in an aqueous solution. The battery lasts far longer than current Lithium-ion models, with the potential to run for almost a decade without charging. Strikingly, the novel chemistry of the battery may also decrease costs of production.

The research published in ACS Energy Letters in February of 2017 describes a flow battery that stores energy in an aqueous solution. According to Wired.com journalist James Temperton, researchers modified the molecular composition of certain elements in the battery including electrolytes, ferrocene and viologen to balance their PH, make them water soluble and stop them from degrading over time. The altered composition of the battery caused two breakthroughs. First, the battery is more neutral and less toxic than traditional batteries. Second, the altered chemical composition of the battery lowers the cost of producing the membrane that separates the chambers of the battery.

The battery has generated buzz within the scientific community as a potential breakthrough in energy storage technology. Many believe that generating aqueous soluble organic electrolytes will be important to producing future batteries with improved life cycles and lower costs. This battery is a large step forward on that front.

With aid from Harvard's Office of Technology Development, the researchers are working with private sector companies to scale the technology for industrial applications. With seemingly endless uses for low-cost, efficient energy storage, the battery has the potential to become a high-impact energy storage solution in the near future.

Beh, Eugene S., et al. "A Neutral pH Aqueous Organic/Organometallic Redox Flow Battery with Extremely High Capacity Retention." ACS Energy Letters (2017). http://www.wired.co.uk/article/flow-battery-energy-grids-harvard-university

Bacteria-Powered Battery Made from Paper
by Justin Wenig

In January of 2017 engineers from The State University of New York at Binghampton demonstrated a bacteria-powered battery on a single sheet of paper that can power disposable electronics. By using a novel manufacturing process, the battery has a lower fabrication costs and takes less time to assemble than previous paper battery models. As a self-sufficient, lightweight device, the battery could be useful to power disposable electronics in dangerous and resource-limited areas such as battlefields.

According to sciencedaily.com, the team of engineers used a sheet of chromatography paper as the base for the battery. On one half of the piece of paper, they placed a small amount of silver nitrate underneath a thin layer of wax to make a cathode. On the other half of the paper, they placed a conductive polymer which served as the anode. After folding the piece of paper together, the engineers added a couple of drops of bacteria-filled liquid. Through cellular respiration, the microbes generated 44.85 microwatts at 105.89 microamps in a 6x6 configuration.

While the amount of energy the battery has generated in initial tests is miniscule compared with the energy output of traditional batteries, it is sufficient to power various life-saving devices such as glucose monitors for diabetics. Most importantly, the battery could be useful because of the wide accessibility of bacteria, from saliva to wastewater, which means that the device is flexible for use in many settings. With a host of potential medical applications on the battlefield and elsewhere, researchers are hopeful they can reduce the cost of the battery and make it commercially available in the near future.

http://onlinelibrary.wiley.com/doi/10.1002/admt.201600194/abstract
https://www.sciencedaily.com/releases/2016/12/161221110606.htm

NON-RENEWABLE ENERGY

Greenhouse Gases

Isotopic Fingerprints of Nitrous Oxide
by Dena Kleemeier

The University of Eastern Finland carried out a study in Northwestern Russia which scientists identified the isotopic fingerprints of nitrous oxide (N_2O) that is produced in Arctic soils. Exceptionally high N_2O emissions were found in Arctic and subarctic regions, which was surprising because N_2O is typically produced by agricultural and tropical rainforest soils, thus scientists thought that concentrations of N_2O in Arctic regions were negligible. The patches with these exceptionally high N_2O emissions originate from "peat circles" which are patches of bare peat surfaces on elevated permafrost peatlands; these peatlands developed through frost action and wind erosion and potentially nitrifier denitrification. Nitrifier denitrification is a microbial process that transforms NH_4 to N_2 through a series of steps (one of which produces N_2O).

This study presents, for the first time, the isotopic fingerprint of N_2O produced by the soils in the Arctic tundra. This opens new opportunities for predicting future trends in atmospheric N_2O as well as identifying climate change migration actions in the Arctic. This is important as the Arctic is a region that is highly susceptible to climate change and N_2O is a powerful greenhouse gas and the second largest contributor to ozone depletion in the stratosphere.

In their conclusion the scientists found that their isotopic data, including SP values that supported the idea that N_2O consumption could be quite large. However their data was not conclusive enough and more laboratory studies are needed to accurately evaluate microbial pathways of production and consumption of N_2O in soils and stable isotope techniques.

Gil, J., T. Pérez, K. Boering, P. J. Martikainen, and C. Biasi (2017), Mechanisms responsible for high N2O emissions from subarctic permafrost peatlands studied via stable isotope techniques, Global Biogeochem. Cycles, 31, 172–189, doi:10.1002/2015GB005370.

Helping Cows Lower Their Methane Emissions
by Dena Kleemeier

This year scientists at Royal Botanic Gardens Kew, Scotland's Rural College (SRUC), and the Senckenberg Biodiversity and Climate Research Centre, Frankfurt, collaborated to publish a paper that illustrates the way in which climate has an affect on grasses. They concluded that plants and grasses growing in warmer conditions are tougher due to their adaptations to heat, preventing heat damage (flower earlier, have thicker leaves, invade new areas). These adaptations provide a lower nutritional value, and are tougher to digest to cows that consume them thus inhibiting the milk and meat yields and directly raising the amount of methane released by the cows. Methane is a greenhouse gas that is 25% more effective at trapping heat in the ozone layer than carbon. When cows consume grass, there are higher levels of methane produced when the grass is more difficult to digest, and over 95% of the methane that they produce is produced from their breath through eructation. Thus the scientists argue that through increased global temperatures there will be adaptations in grasses that will result in higher methane levels being produced from the cows that consume the grasses, a vicious cycle.

Though methane production is generally expected to rise all around the world, the researchers mapped regions where methane will be produced by cows to the greatest extent as result of changing grass quality. These hotspots occur in North America, Central America, Eastern Europe, and Asia. Global meat production has increased rapidly to meet demands and as a result grazing lands have grown to cover 35 million square kilometers, which is a shocking 30% of the earth's surface. The article argues that this immense release of methane can be combatted by cultivation of nutritious plants on grazing lands, however how this will be put in effect was not covered.

Royal Botanic Gardens Kew. "Making cows more environmentally friendly: Research reveals vicious cycle of climate change, cattle diet and rising methane." ScienceDaily. ScienceDaily, 29 March 2017.
    <www.sciencedaily.com/releases/2017/03/170329122607.htm>.

CO_2 Conversion System Converts Greenhouse Gases
by Byron R. Núñez

Sustainable Innovations, Inc. (SI) was awarded a contract from the United States Department of Energy to continue working on its electrochemical process that converts greenhouse gases into usable byproduct. The rising levels of greenhouse gases has increased a demand for new energy solutions that address geopolitical concerns as well as economic ones. Stakeholders, for example, are actively searching for economically viable pathways that can reduce carbon dioxide emissions while developing means to produce fuels that decrease global reliance on oil. This includes, but is not limited to, searching for more efficient ways to utilize traditional fuels such as coal, as well as to capture and recycle the national production of greenhouse gases.

SI's CO2RENEW™ Electrochemical Hydrocarbon Synthesis obtains waste carbon dioxide from smokestacks and converts it to useable chemicals and liquid fuels using electricity from other sources of renewal energy. This system converts waste CO_2 to useful products such as methanol and formic acid via an electrochemical pathway. Methanol fuel is the primary energy carrier in natural gas. CO2RENEW™ produces these hydrocarbons onsite at a fraction of the standard purchase and delivery of the same chemicals. This fuel can then be stored or transported in the existing natural gas pipeline to points of need.

CO2RENEW™ is a promising technology in a time of increasing pressure to develop cleaner, more efficient ways to utilize traditional domestic fuel supplies such as coal. From a political perspective, technologies like CO2RENEW™ and others are desirable since they lessen the national dependence on foreign oil and foreign governments. Ultimately, this process produces a high quality practical fuel that can be used to heat buildings, generate electricity or power vehicles.

PRNewsWire
http://www.prnewswire.com/news-releases/sustainable-innovations-co2-conversion-
    system-converts-greenhouse-gases-and-renewables-into-a-valuable-fuel-with-
    huge-energy-storage-potential-300391823.html
http://www.sustainableinnov.com/

Using Trees and Plants in the Fight Against Climate Change

by Nadja Redmond

We all know that trees take in carbon dioxide and release oxygen, but can this knowledge help the fight the adverse side effects of climate change? Plants have a natural ability to capture carbon dioxide, and several studies are beginning to reveal how this capability could be enhanced to have a maximum impact on carbon dioxide levels in the atmosphere. In 2014, the Intergovernmental Panel on Climate Change announced that plants must be a part of the world's efforts to capture CO_2. On top of their ability to take in CO_2 as they grow, such plants could be designated to be burned and processed into fuels to generate power. Any extra CO_2 produced in the fuel generation would be recaptured. This process is known as "bioenergy plus carbon capture and storage," or BECCS

This process is being tested in several different areas. A new large-scale trial in Illinois will process huge quantities of corn into ethanol, grabbing and storing the 1.1 million tons of CO_2 created in the process underground

There is doubt whether this process will be effective or positive for the environment. At scale, especially the carbon repossession phase, is unproven. Also, the process in its current state doesn't consider how much energy it might take to grow the various plants and the impact of dedicating land to non-food crop, trees, grasses etc. that could be used for food production.

The United Kingdom's Department for Business, Energy, and Industrial Strategy started a $10 million initiative to help investigate how effective BECCS could be, and whether it would be a sustainable approach to locking away CO_2. The project will also seek answers to whether plants and soil can increase their carbon storage.

[https://www.technologyreview.com /s/604260/can-we-fight-climate-change-with-trees-and-grass/].

[https://www.washingtonpost.com/news/energy-environment/wp/2017/04/10/the-quest-to-capture-and-store-carbon-and-slow-climate-change-just-reached-a-new-milestone/?utm_term=.393d22c2bfbc].

[https://www.technologyreview.com/s/604260/can-we-fight-climate-change-with-trees-and-grass/].

Resources: https://www.washingtonpost.com/news/energy-environment/wp/2017/04/10/the-quest-to-capture-and-store-carbon-and-slow-climate-change-just-reached-a-new-milestone/?utm_term=.393d22c2bfbc

https://www.technologyreview.com/s/604260/can-we-fight-climate-change-with-trees-and-grass/

Coal

Indian Firm Pioneers CO_2-to-Baking Soda Carbon Capture Technology

by Lauren Bollinger

An Indian firm has reached a breakthrough in carbon capture technology. Carbon Clean Solutions, a company based in the port city of Tuticorin in Southern India, has developed technology that would take carbon dioxide waste from its coal-fired plant and convert it into baking soda, thereby greatly reducing CO_2 emissions that would otherwise contribute to global warming. The technique is reportedly more efficient than previous methods, requiring less energy and less equipment to run. In total, the plant claims that as much as 66,000 tons of the gas could be captured at the plant each year.

While carbon capture projects are not new, what makes the Tuticorin plant newsworthy is that it utilizes its carbon dioxide waste instead of merely storing it. In comparison, most projects projects aim to bury CO_2 in underground rocks (termed carbon capture and storage or CCS) which reap no economic benefit. The Tuticorin plant by contrast converts its carbon into a usable product, enabling the plant to turn a profit. In fact, the Tuticorin plant represents both the first successful industrial-scale application of carbon capture and utilization (CCU) and the first carbon capture industrial plant running without the use of subsidies.

The Tuticorin process marks a significant step forwards for the clean coal industry, which in the past has proven to be too expensive to implemented widely. Carbon Clean believes that such techniques could ultimately neutralize 5 to 10 percent of the world's CO2 emissions from coal, which account for 65 percent of greenhouse gas emissions around the globe.

Indian firm makes carbon capture breakthrough
https://www.theguardian.com/environment/2017/jan/03/indian-firm-carbon-capture-
breakthrough-carbonclean
A Coal-Fired Power Plant in India Is Turning Carbon Dioxide Into Baking Soda
https://www.technologyreview.com/s/603302/a-coal-fired-power-plant-in-india-is-turning-
carbon-dioxide-into-baking-soda/

Bangladesh to Expand Coal Capabilities to Meet Growing Energy Needs

by Aurora Brachman

Bangladesh has plans to dramatically expand its energy production through coal in the next few years with the assistance of China, Japan, and India. But has not indicated any plans to expand renewable energy development. Other Asian nations have been setting their sights on renewable forms of energy because of an increasingly worsening pollution crisis in the region. The government hopes to expand its use of coal from 2% to 50% of Bangladesh's electricity supply by 2022. There have been vehement protests about this expansion, particularly against a specific plant currently under construction; several people have lost their lives amidst the protests.

Finding new sources of energy is critical to Bangladesh. In 2009, only 47% of the country had access to electricity, but by 2017, 80% of Bangladesh's 160 million people have access to electricity. The country currently relies primarily on natural gas but its reserves are quickly being depleted, thus the interest in shifting to greater reliance on coal. In 2014 the country's finance minister outlined a plan to have 5% of the country's energy come from renewable sources by 2015. As of now, two years past the intended deadline, it has failed to do so.

Although solar has great potential in Bangladesh, it has been largely unsuccessful because of a lack of available space to put the panels. Although, their use has been invaluable in remote and rural areas that otherwise would not have access to electricity.

Because of the affordability of coal, the government feels it has no other option but to pursue its use, but it has not factored in the cost of the environmental and health repercussions. Bangladesh is also uniquely vulnerable to climate change, another sad irony of its decision to expand coal production.

Chowdhury, Kamran R. "By stepping up use of coal, is Bangladesh staring at a grim future?" Scroll.in. N.p., n.d. Web. 14 Feb. 2017. (https://scroll.in/article/828920/by-stepping-up-use-of-coal-is-bangladesh-staring-at-a-grim-future)

Chowdhury, Kamran R. "Bangladesh bets on coal to meet rising energy demand." Climate Home - climate change news. Climate Home, 09 Feb. 2017. Web. 14 Feb. 2017. (http://www.climatechangenews.com/2017/02/09/bangladesh-bets-on-coal-to-meet-rising-energy-demand/)

Coal's Future in the United Kingdom

by Ethan Fukuto

Despite plans set in motion last year to close all coal power plants in Britain by 2025, the continued support for coal by the British government has raised questions and concerns from both the private sector and green energy groups. Continued funding for coal despite the administration's plans come primarily in the form of the capacity market, which provides contracts for energy companies to generate backup energy when supply is low. Scottish Power, one of the United

Kingdom's big six energy companies, and RenewableUK, a non-profit renewable energy trade association, have pointed out the continued subsidies to coal power plants as paradoxical to Britain's no-coal future. Coal power in the UK represented only 12.6% of electricity generation in 2016, compared to 42.9% and 22.6% for gas and nuclear, respectively. Steven Holliday, formally the chief executive of the National Grid which runs the capacity market, noted that the UK's future power grid will be more flexible to support increases in demand. In November of 2016, the British government promised £730 million in support for renewable energy. Increased use of smart-meters in homes and businesses, as well as battery storage, points to further decreases in the use of coal. Even without government intervention, coal plants are expected to shut down by 2022 due to air pollution and climate change laws. Regardless, coal—along with biomass plants—received 20% of contracts in the capacity market auction on February 3 to provide backup power for next winter, second only to gas which received 40% of contracts. Scottish Power has called for tighter emissions standards that must be met to be considered for future auctions (the next one being in December 2017), one which coal is not likely to meet. For the time being, Scottish Power and RenewableUK are critical of the government's allocation of tax money to support a dying subsection of the energy sector.

Vaughan, Adam. "Ban coal from backup power subsidy scheme, says Scottish Power". The Guardian. 30 January 2017.
https://www.theguardian.com/business/2017/jan/30/ban-coal-backup-power-subsidy-scheme-scottishpower

Vaughan. Adam. "Britain's last coal power plants to close by 2025". The Guardian. 9 November 2016.
https://www.theguardian.com/environment/2016/nov/09/britains-last-coal-power-plants-to-close-by-2025

Williams, Diarmaid. "Auction contracts announced for UK power plants". Power Engineering International. 6 February 2017.
http://www.powerengineeringint.com/articles/2017/02/auction-contracts-announced-for-uk-power-plants.html

Navajo Generating Station Will Close Despite President Trump's Interference

by Genna Gores

Three years ago the EPA struck a deal with the owners of the Navajo Generating Station, a coal-fired power plant, to shut down its operation by 2044. However, due to the shifting economy in energy it will now close in 2019. The Navajo Generating station is the largest coal plant in the western U.S., and is the seventh largest, individual source of climate pollution emitting 14 million metric tons of CO_2 per year. Closing this plant could potentially save $127 million in health costs, but also has many societal and economic implications for Arizona and the U.S.

The coal market is currently in decline, and many other plants besides the Navajo Generating Station are closing too. Despite

President Trump's plan to boost coal and move away from renewables, the current energy market will not change. New policies supporting coal-fired power, eliminating environmental protections, and hindering the development of renewables will only slow down the downward trend of coal. Over the past few years, the energy market has had a huge increase in supply from renewables and natural gas but the demand for energy has stayed relatively the same. Since coal is more expensive and environmentally destructive, simple economics will force this form of energy out of the market. The plant is already losing costumers, like the Los Angeles and Nevada Electric utilities, because of various state emission standards; so at this point the plant is not as economically feasible as envisioned when a 2044 decommissioning was contemplated.

The Navajo Generating Plant preemptive closure has many positive effects, but it will be a tough transition for the Navajo and Hopi people who are employed by the plant. While the tribes knew the plant was going to close, it ishappening much quicker than expected, and the tribes had not planned to replace the hundreds of jobs provided by the plant. These tribes want President Trump to intervene, but at this point his policies cannot change the outcome. For these tribes, though, there is the potential for better jobs in the future. The site of the plant is one of the best areas for solar development in the U.S. and this could be a new opportunity for healthier and more environmentally friendly jobs. [http://www.latimes.com/opinion/readersreact/la-ol-le-navajo-generating-station-20170225-story.html]

Rebranding the Coal Industry

by Nadja Redmond

Due in part to the current administration's preference towards non-renewable energy sources and the possibly detrimental reversal of the Climate Action Plan, the three largest American coal producers are stepping up to attempt to remedy negative perceptions of the coal industry. Executives from these companies aim to rebrand the image of the struggling coal industry as a contributor, not a road block, to a future focused on clean energy. The leadership at Cloud Peak Energy, Peabody Energy, and Arch Coal is starting to voice larger concern about greenhouse gas emissions, for example. The executives of these companies are going further in their mission as well; in some instances, they are directly engaging and making common cause with their most adamant critics, such as the Natural Resources Defense Council and the Clean Air Task Force. They've joined forces currently to lobby for a tax bill that would expand government subsidies to reduce the environmental impact of coal burning.

The technology they hope to implement and improve through this effort would aid in carbon capture and sequestration, an immature technology at the moment. Richard Reavey, vice president of Cloud Peak Energy, had this to say regarding the change in tone: "We have to accept that there are reasonable concerns about carbon dioxide and climate, and something has to be done about it. It's a political reality, it's a social reality, and it has to be dealt with". The "social reality" he is in part talking about connects to the main argument these industry leaders are making for continued support of coal and other fossil fuels; the executives say that renewable energy sources are not sufficient to meet energy needs in coming years. They believe coal and other fossil fuels will continue to lead the industry, which is why they advocate for reform and support of the new technologies they propose, rather than rejection

http://nyti.ms/2msUruC.
[http://nyti.ms/2msUruC].
[https://www.fastcoexist.com/3068458/the-largest-coal-plant-in-the-western-us-is-closing-decades-ahead-of-schedule]
[http://nyti.ms/2msUruC].

Natural Gas

Flaring Natural Gas

by *Emily Audet*

Natural gas flaring involves burning natural gas at oil and gas extraction sites. Pipes feed gas to flare stacks, where the gas is burned. Flaring can last from days to weeks and produces large quantities of heat and noise. Gas companies flare for a number of reasons, including to test completed wells, to release pressure in wells for safety reasons, and during processing.

In February 2016, the Bureau of Land Management (BLM), which has authority over federally-owned property, proposed rules to decrease flaring natural gas. The regulation would require oil and gas companies to use best available technologies to avoid flaring, to perform routine evaluations of sites for leaks, and to improve used technologies to minimize the venting of natural gas into the air. The regulation also defines when oil and gas companies are required to pay royalties for flaring. The rationale behind these regulations asserts the necessity to preserve natural gas as a resource, as flaring natural gas releases it into the atmosphere without putting it to use.

As of February 2017, reporters have predicted that the new administration and Congress will push to halt these regulations as part of their efforts to deregulate industry. In a February 2017 article, Bret Wells, a University of Houston Energy Fellow, argues in support of moving forward with the proposed BLM regulations on flaring. Wells argues that preserving as much natural gas as possible (ie. through not flaring natural gas) aids the U.S.'s energy production and contributes to the country's energy independence.

Flaring natural gas releases methane, and this issue remains one of the primary concerns related to the process. BLM observed that methane has 25 times the effect on global warming than carbon dioxide. The best approach from an environmental and long-term national economic perspective is to recover all gas. If gas cannot be captured, however, flaring the gas is more environmentally sound that directly releasing the gas into the atmosphere. Green completions, also known as reduced emission completions, can be used at hydraulic fracturing wells as a best practice. Green completions separate gas from water as they rise to the soil level—this process allows the extra gas to be captured and sold, rather than having to be flared.

http://www.forbes.com/sites/uhenergy/2017/02/03/when-flaring-natural-gas-becomes-
   political-needless-regulation-or-good-conservation/ - 5ef4a261523f].
[https://www.blm.gov/wo/st/en/prog/energy/oil_and_gas/methane_and_waste.html;
[http://www.epa.state.oh.us/portals/27/oil and gas/basics of gas flaring.pdf].
Bureau of Land Management. "Proposed Methane and Waste Prevention Rule." N.p., 4–25
   2016. Web. 7 Feb. 2017.
   https://www.blm.gov/wo/st/en/prog/energy/oil_and_gas/methane_and_waste.h
   tml
Ohio EPA. "Understanding the Basics of Gas Flaring." Nov. 2014. Web. 6 Feb. 2017.
   http://www.epa.state.oh.us/portals/27/oil and gas/basics of gas flaring.pdf
Wells, Bret. "When Flaring Natural Gas Becomes Political -- Needless Regulation Or Good
   Conservation?" Forbes. N.p., 3 Feb. 2017. Web. 7 Feb. 2017.
   http://www.forbes.com/sites/uhenergy/2017/02/03/when-flaring-natural-gas-
   becomes-political-needless-regulation-or-good-conservation/ - 5ef4a261523f

New Study Points to Malfunctions in Production Process as a Cause for "Super-Emitters"

by Emily Audet

A 2015 study by the Environmental Defense Fund found that only 10% of natural gas production sites comprise 80% of all methane emissions, adding credence to the idea that a minority of "super-emitters" produce the majority of emissions from natural gas production. A 2016 Stanford study found similar results to confirm the existence of super-emitters: 5% of methane leaks comprise 50% of all methane leakage volume. Focusing on regulating super-emitters, therefore, poses both a large challenge and possible high payoff to decreasing methane emissions. Adam Brandt, assistant professor of energy resources engineering at Stanford, suggests that focusing on preventing super-emissions can be cost-effective, as less expensive and less sensitive monitoring equipment can be used to detect methane leaks.

In a follow-up paper published on January 16, 2017, the Environmental Defense Fund tried to find the cause for disproportionately high emissions on these sites. The research compared measurements of overall emission levels in the air with measurements of emissions from specific site technologies in the Barnett Shale region of Texas. The study found that estimating methane emissions through examining the individual emissions of each piece of equipment underestimated the actual emissions by about one-third. Researchers concluded that malfunctions in the production process cause this discrepancy and are therefore responsible for creating "super-emitters." These malfunctions are not limited to issues with on-site technology but rather include poor functioning in the entire production process.

In a blog post summarizing the paper, Daniel Zavala-Araiza, the paper's primary author, points to two more conclusions. First, when and from which sites high amounts of methane are being released—in other words, which sites are super-emitters and when—is dynamic.

Second, Zavala-Araiza recommends increased evaluation of natural gas production sites to prevent the super-emitter phenomenon. Because different sites are super-emitters at different times, he recommends increased monitoring as the best solution to identify and prevent disproportionately high methane emissions.

The results of this paper found significantly higher emission levels than those measured by the EPA, since the EPA's Inventory uses predicted emissions from site technologies and process. Zavala-Araiza asserts that while these results were obtained in the Barnett Shale area, they most likely reflect similar patterns of super-emitters across the country, and even the world.

[http://www.nature.com/articles/ncomms14012].
[http://news.stanford.edu/2016/10/26/super-emitters-responsible-bulk-u-s-methane-emissions/].
[http://breakingenergy.com/2017/01/20/super-emitters-are-real-here-are-three-things-we-know/].
Than, Ker. "'Super Emitters' Responsible for Most U.S. Methane Emissions." Stanford News. N.p., 26–26 Oct. 2016. Web. 14 Feb. 2017. http://news.stanford.edu/2016/10/26/super-emitters-responsible-bulk-u-s-methane-emissions/
Zavala-Araiza, Daniel. "Super-Emitters Are Real: Here Are Three Things We Know." Breaking Energy. Environmental Defense Fund Energy Exchange Blog, 20 Jan. 2017. Web. 14 Feb. 2017. http://breakingenergy.com/2017/01/20/super-emitters-are-real-here-are-three-things-we-know/
Zavala-Araiza, Daniel et al. "Super-Emitters in Natural Gas Infrastructure Are Caused by Abnormal Process Conditions." Nature Communications 8 (2017): 14012. Web. 14 Feb. 2017. http://www.nature.com/articles/ncomms14012
TAGS: Environmental Defense Fund, Nature Communications, Daniel Zavala-Araiza, super-emitters, methane, emissions, natural gas, Stanford, Ker Than
https://www.linkedin.com/company/environmental-defense, @EnvDefenseFund, kerthan@stanford.edu

Clean Coal faces Competition from Clean Natural Gas

by Kieran McVeigh

A top priority through early 2017 for president Trumps has been putting coal miners back to work, and one approach is to use clean coal technology. The technology president Trump refers to when he talks about clean coal is a process to capture some of the carbon dioxide that is released when burning coal. This recaptured carbon dioxide can be pumped back under ground where it makes existing oil refineries more productive. Although this technology has some promise it also has drawbacks, including recapture rates of as low as 10% of the carbon dioxide emissions of the burn coal, which ends up making clean coal not so clean.

While the technology for clean coal may have some flaws, it also has a big problem in competition coming from the natural gas industry. A new natural gas company in Texas, NET power has created a test power plant that uses clean natural gas technology to generate power. Where this company is a little bit different, is that rather then using steam to power their energy-generating turbine they use the carbon dioxide from the burned natural gas. Much of this carbon dioxide is then recycled back through the turbine for maximum efficiency, and the carbon dioxide that is released is recaptured in much the same way that the carbon dioxide emissions from "clean" coal are.

NET power bills titself as "clean" and environmentally friendly; others are not so sure, arguing because NET power removes the natural gas from the ground it isn't renewable, and therefore not environmentally friendly. While if you listen to NET power talk it seems as if their plant will have zero emissions, but low rates of carbon dioxide capture for clean coal suggest that is might not be quite as environmentally friendly as advertised.

Although both clean coal and clean natural gas offer potential promise they both face a big issue economically; solar and wind power are getting cheaper and cheaper and although they may not be a priority for president Trump, they spell bad news for clean gas and coal companies looking for investors.

Joyce, C. (2017, April 10). Natural Gas Plant Makes A Play For Coal's Market, Using 'Clean' Technology. Retrieved April 11, 2017, from http://www.npr.org/2017/04/07/522662776/natural-gas-plant-makes-a-play-for-coals-market-using-clean-technology

Brady, J. (2017, March 29). Climate-Friendly Coal Technology Works But Is Proving Difficult To Scale Up. Retrieved April 11, 2017, from http://www.npr.org/2017/03/29/521926674/climate-friendly-coal-technology-works-but-is-proving-difficult-to-scale-up

Methane Leaks: An Unnatural Byproduct of Natural Gas Production

by Justin Wenig

While many individuals think of carbon dioxide emissions as the primary cause of climate change, methane emissions are also a major contributor. In 2014, the Environmental Protection Agency (EPA) estimated that methane accounted for 11% of U.S greenhouse gas emissions. Strikingly, a 2016 review by the EPA found that more than half of U.S methane emissions stemmed from methane leaks resulting from natural gas production chains. Clearly, as energy production in the United States continues to shift from coal to natural gas, preventing methane leaks is a key to deterring climate change.

In 2015, the EPA began requiring producers of natural gas to regularly scan for and fix methane leaks. However, with new president Donald Trump's administration skeptical of EPA regulation, it is possible that this requirement could be rolled back. Since detecting methane leaks is a labor and capital-intensive process, if the EPA regulation is overturned, companies with tight profit margins or an appetite for risk could stop looking for methane leaks altogether. In turn, U.S emissions would rise and the rate of climate change would accelerate. Even more jarringly, recent public pressure levied on energy producers to invest in natural gas as opposed to coal might not reduce emissions at all.

However, there is hope from the private sector that natural gas producers could continue to scan for methane leaks if EPA regulation forcing them to do so is overturned. In October of 2016, General Electric unveiled a helicopter drone called "Raven" that is designed to detect methane leaks cost-effectively. While under the current method, a worker must circle the natural gas well with an infrared camera to detect leaks, the Raven utilizes cutting-edge laser detection technology to pinpoint leak location and magnitude. Here is to hoping that the Raven or another private sector innovation will not only help the US avoid the harmful environmental consequences of a potential rollback in EPA regulations, but also end emissions from methane leaks altogether.

https://www.epa.gov/ghgemissions/overview-greenhouse-gases

https://www3.epa.gov/climatechange/Downloads/ghgemissions/US-GHG-Inventory-2016-Main-Text.pdf

http://www.geglobalresearch.com/news/press-releases/ge-opens-new-oil-gas-rd-center-in-oklahoma-showcases-smart-sensing-drone-advanced-labs-and-emerging-digital-technology

Petroleum

IEA Predicts Long-Term Increase in Oil Demand and Corresponding Price Rise

by Emily Audet

Oil 2017, a market forecast published by International Energy Agency (IEA), predicts that after 2020, global oil demand will quickly outpace supply, which would likely cause an increase in oil prices From 2017, global oil demand is predicted to increase by about 1.2 million barrels per day (bpd) until 2022.

Increasing demand for oil among middle- and low-income countries, especially those in Asia, like India and China, is driving this increased global demand. On the supply side, investment in oil fell to record lows in 2015 and 2016. The report predicts that oil supply will increase in the U.S., Canada, and Brazil, among other countries, until 2020, but by 2020, this supply-side growth will halt if investment rates remain low.

In 2016, IEA published a report called World Energy Outlook, which made predictions about oil demand trends going into the future. The IEA predicted that global oil demand would increase from 92.5 million bpd in 2015 to 117 bpd or 103 bpd by 2040, depending on how many countries pass regulations trying to decrease oil demand, which corresponds to a compound rate of increase of 1% or 0.5% per year, respectively. The global number of electric vehicles is predicted to increase from 1.3 million in 2015 to 15 million by 2022. The IEA, however, expects this increase in electric vehicles to only remove 200,000 bpd from the global oil demand until 2022. In order for atmospheric carbon levels to fall to 450 parts per million, global oil demand would have to shrink to 73 bpd by 2040.

Interestingly, among a decrease in global oil investment over the last few years, investment in U.S. shale oil has been increasing. The IEA expects the U.S. to produce most of the new supply over the coming years. The U.S. oil supply is also more sensitive to price changes compared to other countries. IEA predicts that Iraq, Iran, and the United Arab Emirates will provide most of the new supply within the Organization of the Petroleum Exporting Countries (OPEC).

[https://www.ft.com/content/6506aba2-0265-11e7-ace0-1ce02ef0def9].
[https://www.forbes.com/sites/rrapier/2016/11/30/iea-projects-a-75-increase-in-oil-
    prices-by-2020/ - 49727c67a675].

[http://www.iea.org/newsroom/news/2017/march/global-oil-supply-to-lag-demand-after-2020-unless-new-investments-are-approved-so.html].

Rapier, Robert. "IEA Projects a 75% Increase In Oil Prices By 2020." Forbes. N.p., 30 Nov. 2016. Web. 7 Mar. 2017. https://www.forbes.com/sites/rrapier/2016/11/30/iea-projects-a-75-increase-in-oil-prices-by-2020/#49727c67a675

Raval, Anjli, and Gregory Meyer. "IEA: Oil Investment Drought Threatens Price Surge." Financial Times. N.p., 6 Mar. 2017. Web. 7 Mar. 2017. https://www.ft.com/content/6506aba2-0265-11e7-ace0-1ce02ef0def9

"Global Oil Supply to Lag Demand after 2020 Unless New Investments Are Approved so." International Energy Agency. N.p., 6 Mar. 2017. Web. 7 Mar. 2017. http://www.iea.org/newsroom/news/2017/march/global-oil-supply-to-lag-demand-after-2020-unless-new-investments-are-approved-so.html

Trump Moves to Continue Construction of the Dakota Access Oil Pipeline

by Lauren Bollinger

On his second weekday in office, President Trump filed an executive order to reopen construction of the Dakota Access Pipeline, a project that was formerly blocked by the Obama administration after months of protests by Native American activists. The 1,172 mile-long pipeline is slated to run through four states, from the Bakken oil fields of North Dakota through Illinois, and transport around half a million (470,000) barrels of crude oil per day.

Originally planned for delivery by January 1st of this year, the project was stalled after widespread protests led by Native American activists which gained international attention. In the early stages of planning, the pipeline was proposed to run through Bismarck, North Dakota, but was rerouted to run adjacent to the Standing Rock Reservation, after concerns from Bismarck residents. Members of the Standing Rock Sioux oppose the project as they argue it threatens environmental safety and indigenous sovereignty, as the pipeline is slated to run only a mile from their tribal borders.

It is worth noting that Trump is a former stakeholder in two of the companies behind the Dakota Access Pipeline project, with nearly $1 million invested in Energy Transfer Partners and $250,000 in Phillips 66 in 2015, but has since sold his shares as of summer 2016.

This move marks the first major step by Trump to reverse his predecessor's environmental policies in favor of his America First Energy Plan, which advocates for energy independence. In addition to this measure, Trump administration has signed actions to resume construction of the Keystone XL pipeline, a similar project that was gutted by Obama in 2015, and expedite environmental reviews for high-priority infrastructure projects.

"Trump reopens door to building Keystone XL and Dakota Access pipelines." http://www.latimes.com/politics/washington/la-na-trailguide-updates-trump-opens-keystone-xl-and-dakota-1485275942-htmlstory.html

"Trump signs executive actions to advance the construction of the Keystone XL and Dakota Access oil pipelines" https://apnews.com/45a8e5feb306444a8a180f6f3e6d7a97

"Trump signs actions to advance Keystone XL, Dakota Access Pipelines"
    http://abcnews.go.com/Politics/trump-advances-keystone-xl-dakota-access-
    pipelines/story?id=45008003

ADNOC's Push for Partnerships

by Dominique Curtis

Abu Dhabi's National Oil Company (ADNOC) has been busy the past few months building partnerships, finding investors, and making deals. ADNOC took a big hit when oil prices collapsed in 2014. Now new leaders have joined ADNOC with new ideas and plans to help ADNOC rise again. Mr. Sultan Al Jaber became the chief executive of ADNOC over a year ago. Al Jaber is also Minister of State and a member of the Supreme Petroleum Council, which is The Emirate's top oil sector decision-making body (McAuley, 2017). Al Jaber says that Abu Dhabi will more than triple its domestic petrochemical output by 2025 and it plans to expand overseas while looking to develop other supplies of gas. Al Jaber is in search of more long term partnerships than ADNOC has had in the past.

Multiple companies have responded to Al Jabe's call for partnerships. ADNOC signed a $1.77 billion dollar deal with China National Petroleum Company (CNPC) for 8 % of ADNOC's onshore oilfield concession. This billion dollar deal includes a 40 year partnership and about 15 oilfields. The day after this deal was signed China Energy (CEFS) made a deal for 4 % of ADNOC's stakes. These deals help Al Jaber meet his goal for 40 percent foreign stakeholders.

Now Abu Dhabi and its biggest competitor, Saudi Arabia are coming together to enhance their oil and gas renewables. ADNOC and Saudi Aramco will work together to develop new technologies and share knowledge about clean energy. Mr. Al Jaber says that these different approaches to partnerships will help ADNOC to meet its ambitious targets for its five-year business plan and the broader 2030 goals (Graves, 2017).

Graves, LeAnne. "Abu Dhabi and Saudi Arabia to Team up on Energy Technology and
    Renewables." The National. N.p., 02 Apr. 2017. Web. 03 Apr. 2017.
Graves, LeAnne. "Adnoc Awards China's CNPC 8 per Cent of Adco Concession." The
    National. N.p., 19 Feb. 2017. Web. 04 Apr. 2017.
McAuley, Anthony. "Adnoc Chief Targets Downstream Partnerships amid Gas Glut." The
    National. N.p., 04 Apr. 2017. Web. 03 Apr. 2017.
Mills, Robin. "Robin Mills: Abu Dhabi Oil Agreements with China Continue Shift towards
    New Partners." The National. N.p., 05 Mar. 2017. Web. 03 Apr. 2017.

Oil Giants Drill near Amazon Reef

by Dena Kleemeier

This past year a reef was reportedly discovered at the base of the Amazon River. The reef is 600 miles long and extends from the southern tip of French Guinea to the Brazilian Maranhão state, and was left undiscovered for years as it lies under a thick plume of muddy water where the Amazon River pours into the Atlantic. The plume of mud shields the reef from sunlight, however the reef continues to thrive as it generates energy using chemosynthesis rather than photosynthesis. Scientists who discovered the reef say that it is home to species that haven't been found anywhere else, such as the Amazon River dolphin, manatee, and hawkbill sea turtle. This reef is incredibly important to the biodiversity of the area and has huge potential for new species and biotechnology. Additionally the fish feed the fisheries positively affecting the socioeconomic status of locals.

However the oil giants BP and Total own 5 deep-water exploration licenses in the Foz do Amozanas just 8 kilometers from the reef, and plan to start drilling this year. Although the companies previously did environmental assessments to analyze the risks of the exploration, they were conducted previous to the discovery of the reef. A spokesperson from BP argued that the reef was discovered in the 1970s and thus it was common knowledge whilst the companies were conducting their environmental studies to analyze risks. However, this is simply untrue; the reef was discovered in 2014 and wasn't announced to the public until 2016. Ignoring the possible environmental impacts, BP and Total plan to continue with their plan to drill after IBAMA (the Brazilian government offshore oil/ gas regulator) assesses the environmental impacts and gives permits to begin exploration drilling.

So, even though Total is "well aware of the reef, and understand the need to preserve it" they still won't take the time or money to make sure that their actions won't have detrimental effects on the reef. I guess that's capitalism for you.

Carter, Lawrence. "Amazon reef: BP and Total set to drill for oil near newly discovered coral reef." Energydesk. N.p., 30 Jan. 2017. Web. 28 Mar. 2017.

DAPL Update: The Oil Is Flowing… How Long 'Til It Spills?

by Genevieve Kules

The Dakota Access Pipeline, a crude oil transportation facility stretching from North Dakota to Illinois, has been fought from its inception, with increased publicity in the past year. Though it was slowed and halted for about three months, oil has entered the system in a reservoir under the Missouri River in North Dakota. It may be a blow to the water protectors, but the fight is not over. The Standing Rock and Cheyenne River Sioux Tribes have lawsuits against DAPL that aim to divert it from sacred and tribal lands, if not shut it down entirely on the basis of unsafe drinking water.

The construction under Lake Oahe was the final piece to be built. Drilling under the Missouri River, a resource that provides drinking water to millions of people not only the Standing Rock and Cheyenne River Sioux Tribes, was stopped by President Obama and the Army Corps of Engineers in late 2016 after much advocacy and hard work on the part of water protectors and members of those tribes. Passing oil underneath the Missouri River has too many high risk factors that endanger the earth, water, humans, and animals to move forward. The rightful protectors of sacred lands prohibit the pipeline from passing through their territories. That prohibition is not being respected by the US government or the corporate entities responsible for the pipeline.

Drilling has already caused large destruction, harm, and emotional trauma for persons connected to the land. The risk of more harm, trauma, and loss of land and rights to govern one's own land is unacceptable. In my opinion, the US's extreme focus on capitalism and money has created an apathy for the environment and the original protectors of the land. According to the water protectors, the flowing of oil through this pipeline marks a terrifying day, but does not stop - and perhaps strengthens - the fight against it.

Cote, Rachel Vorona. "Dakota Access Pipeline Has Been Filled With Oil And Prepared for Service." Jezebel. Gizmodo Media Group, 26 Mar. 2017. Web. 27 Mar. 2017. <http://jezebel.com/dakota-access-pipeline-has-been-filled-with-oil-and-pre-1793698685>.

"Oil Placed in Dakota Access Pipeline as Service Set to Begin." RT International. TV-Novosti, 28 Mar. 2017. Web. 28 Mar. 2017. <https://www.rt.com/usa/382509-dapl-oil-service-prep/>.

Newcomb, Steven. "Indigenous Sovereignty and the Political Subordination of Our Nations."Indian Country Media Network. Indian Country Today Media Network, 24 Mar. 2017. Web. 28 Mar. 2017. <https://indiancountrymedianetwork.com/news/opinions/indigenous-sovereignty-political-subordination-nations/>.

Canada's Expansion of Oil Pipelines
by Sara R. Roschdi

Canada and Prime Minister John Trudeau are known for their social welfare programs and their liberal politics. The country is also taking steps backwards in addressing climate change and the environment by expanding fracking and oil pipelines. This supports the country's multinational corporations at the extent of the indigenous people who are resisting to protect their ancestral homelands, our worlds clean water supplies and the ecosystem. In September 2016, over 50 Indigenous nations across North America came together to sign a historic pan-continental treaty against the expansion of pipelines into their territories and ancestral homelands. This treaty called the Treaty Against Tar Sands Expansion brought together First Nation peoples to protect their water supplies and the mother earth.This alliance against oilsands opposes the building of pipelines, tankers, and rail projects in the region including the TransCanada's Energy East Pipeline, Kinder Morgan's Trans Mountain expansion, Enbridge's Line 3 pipelines, and Enbridge Northern Gateway (McSheffrey, 2016). Prime Minister Justin Trudeau went against this treaty and announced the expansion of the Trans Mountain pipeline, the Enbridge Line 3 pipeline and the continued extraction of oil from the Alberta's tar sands. The Guardian Reports that Canada is only 0.5% of the world's population but is planning to sell a third worth of the earth's carbon budget set. The Canadian diplomat Catherine McKenna called for stricter enforcement and higher expectations at the Paris climate talks on climate change but the country is extracting and selling oil in a way that counteracts this talk about the nation taking major steps to combat climate change. It is contradictory to boast of the reduction of Canada's carbon emission and eco footprint when they are fracking and facilitating other nations and multinational corporations increased oil consumption and thus increasing global CO_2 emissions. This expansion of oil pipelines benefits multinational corporations and is a dangerous threat for indigenous peoples and global climate change.

(https://www.theguardian.com/commentisfree/2017/apr/17/stop-swooning-justin-trudeau-man-disaster-planet?CMP=fb_gu) .

(https://warriorpublications.wordpress.com/2016/09/22/first-nations-across-north-america-sign-treaty-alliance-against-the-oilsands/#more-11929) .

(http://www.huffingtonpost.ca/2016/11/29/trudeau-pipeline-approvals-reaction_n_13311080.html).

https://www.theguardian.com/commentisfree/2017/apr/17/stop-swooning-justin-trudeau-man-disaster-planet?CMP=fb_gu

https://warriorpublications.wordpress.com/2016/09/22/first-nations-across-north-america-sign-treaty-alliance-against-the-oilsands/#more-11929

http://www.huffingtonpost.ca/2016/11/29/trudeau-pipeline-approvals-reaction_n_13311080.html

Hydraulic Fracturing\Fracking

Shale Gas Fracking causing friction in the UK

by Dominique Curtis

Controversy in the UK community has sparked over shale gas. Whitmarsh (2015) discusses how shale gas is the newest project the UK government has suggested to help reduce their reliance on energy ports. The community has questioned the UK's method of fracking to extract the shale gas because fracking is known to use large amounts of water and the chemicals used in the process are toxic. Researchers and the UK government have tried to explain the great benefits that shale gas will have on the economy and the environment while attempting to pacify the communities' concerns. Environmental groups still protested about how fracking will contaminate and decrease the availability of water supply, and cause erosion and changes in the temperature of the water in aquatic habitats. The responses to these concerns were rebuttals about how these side effects would be minimal and that all energy technologies have unwanted side effects. The Prime Minister exclaimed how shale gas will drive down energy cost and create more jobs. Although it is true that there would be more jobs, the increase wouldn't be any more significant than jobs for other energy sources. Also, shale gas production would not reduce energy cost of the community because of the UK's place in the natural gas market. Researchers have noted other questionable facts that leave room for skepticism and concerns about the government's investment and interest in shale gas. The International Energy Agency (IEA) preached about how shale gas will contribute to 14% of global production by the year 2035 and will create a low carbon future in the year 2050. It is true that shale gas carbon footprint would be lower than coal-fired power but shale gas still has a higher carbon footprint than conventional gas, nuclear power, and renewable gases. So the big question is who really benefits from shale gas? Hammond explains that "UK overall might benefit from improved energy security and reduced balance of payments but local communities bear the environmental and health risk of fracking (Hammond, 2017)." Some of the community members are 100% against shale gas fracking, some of the community members are okay with fracking as long as it's not in their backyard, but for the most part many UK community members are unaware and uneducated on shale gas fracking. shale gas fracking in the UK is still in the early stages and there are still many uncertainties and lack of awareness about the benefits and the risks.

Whitmarsh, Lorraine, Nick Nash, Paul Upham, Alyson Lloyd, James P. Verdon, and J.-Michael Kendall. "UK Public Perceptions of Shale Gas Hydraulic Fracturing: The Role of Audience, Message and Contextual Factors on Risk Perceptions and Policy Support." Applied Energy 160 (2015): 419-30. Web.

Hammond, Geoffrey P., and Áine Oâ€™Grady. "Indicative Energy Technology Assessment of UK Shale Gas Extraction." Applied Energy 185 (2017): 1907-918. Web.

New Report Spills on Fracking Leaks

by Siena Hacker

A February 2017 study in *Environmental Science & Technology* published data exposing 6,648 fracking-related spills over the past ten years in just four states. A February 2017 ThinkProgress article reports that the study, which was funded by the Science for Nature and People Partnership (SNAP), investigated spills in Colorado, New Mexico, North Dakota, and Pennsylvania. Across the four states, 50% of the spills were related to the storing of oil and gas in flowlines, which are pipelines that connect wells to processing sites. According to a February 2017 ResearchGate article, more than 75% of well spills occurred within the first three years of the well's operation. The article explains that this higher incident rate is likely because the most drilling, hydraulic fracturing, and production occur during this time. However, studies should not be limited to these first three years as they would miss crucial subsequent spills, especially transport leaks. A prior EPA study of eight states, solely focused on the fracturing stage, found just 457 spills between 2006 and 2012. However, this EPA study ignored the potential for spills to occur later in the lifecycle of the well. This reflects a larger problem in analyzing fracking spills data—namely that national, uniform reporting regulations do not exist. The February 2017 SNAP study thus supports the creation of stronger, more consistent industry standards for reporting spills. Laura Patterson, a Duke University water policy specialist and contributing author to the study, recommended creating a uniform set of requirements across states to facilitate the aggregation and comparison of data for the benefit of well owners and the government. According to Patterson, standardizing spill data and making it more accessible could inform stakeholders on where to direct efforts to prevent future spills. Additionally, states could utilize the data to create regulations that target the conditions and locations correlated with spills. According to Hannah Wiseman, another author of the report, the results of the study indicate that the majority of spills are preventable. Understanding how to prevent these spills could save well owners money and help state governments mitigate harmful environmental impacts.

ResearchGate: https://www.researchgate.net/blog/post/study-of-fracking-in-four-states-uncovers-over-6600-spills

ThinkProgress: https://thinkprogress.org/6600-spills-fracking-four-states-92df12858f6e#.1anv9u1zj

"Unconventional Oil and Gas Spills: Risks, Mitigation Priorities, and State Reporting Requirements" http://pubs.acs.org/doi/abs/10.1021/acs.est.6b05749

Nuclear Power

China's Nuclear History

by Cybele Kappos

An inactive Chinese nuclear site is currently being repurposed for tourism in order to honor the efforts of the soldiers who created the site. Qin's 2017 article explores the beginning of the project in the 1960s and the circumstances surrounding it as well as the mixed reactions to its construction. The project, titled 816, was an effort to build a nuclear reactor without interference of the Soviets. To reduce the risk of an attack, the Chinese decided to build the project underground. What resulted was the world's largest artificial cave and an impressively resilient plant that could supposedly handle several thousands of tons of explosives as well as a magnitude 8 earthquake. The project is a replica of the Soviet-designed 404 project in the Gansu province. In the 1980s, the construction of the plant was nearly complete when it was brought to a halt for political reasons. Although the plant served as a chemical fertilizer factory, there were several concerns about forgetting about the history of the plant. These involved people who worked on the excavation and construction of the plant saying that this project took a lot out of the workers and even a few lives and that to forget would dishonor these people's efforts. In order to respond to these concerns, the plant underwent renovations and became a tourist attraction. Visitors can only see about one-third of the cave which contains an astonishing 13 miles of tunnel roads. Qin quotes one of the tour guides saying that the plant is not only a monument to Chinese communism but is also an important part in the development of China's national defense and nuclear program. Others argue that the decision to halt the project was a sound one since it would not have contributed much to China's nuclear program. Some also seem disappointed that this project, which took a lot of effort by the soldiers, ultimately became a tourist site.

Qin, Amy. A Chinese Nuclear Site, Hidden in a Mountain, Is Reborn as a Tourist Draw. Jan 24, 2017

https://www.nytimes.com/2017/01/24/world/asia/china-fuling-nuclear-816.html?rref=collection%2Ftimestopic%2FNuclear%20Energy&action=click&contentCollection=energy-environment&region=stream&module=stream_unit&version=latest&contentPlacement=1&pgtype=collection&_r=0

http://www.upi.com/China-reopens-former-underground-nuclear-facility-to-the-public/4071474912467/

NuScale Submits Miniature Nuclear Reactor Design

by Kieran McVeigh

Nuclear power once hailed as the power of the future, has lost much of its hype with the advent of other renewable sources that do not have such nasty potential side effects. NuScale, a new Nuclear power company is aiming to change this by rethinking the design of nuclear power plants, which are currently hulking structures made out of many tons of concrete and steel. The aptly named NuScale has submitted a design for what it describes as a modular Nuclear power plant, to the Nuclear Regulatory commission for review. Each "Module," a prefabricated 50-megawatt reactor, is drastically smaller than existing designs, small enough to fit on a flatbed truck. Since each reactor is so small, it is unlikely that there is enough uranium to cause a catastrophic nuclear meltdown.

NuScale is marketing these reactors as a way to cut down on the upfront costs of nuclear power, saying their design (assuming it is approved) will make nuclear power plants more affordable. This reduction of costs, is achieved by prefabricating each reactor and then sending it to the construction site. Once there, multiple reactors can be linked to create a larger nuclear power plant. NuScale claims that its design will limit operation costs, as fewer people are required to keep the plants running then traditional plant designs. Opponents argue that because of the smaller reactors the power produced will actually be more expensive, which the company acknowledges may be true, but emphasizes by that this will be offset by decreased construction costs.

Another potential use for these reactors is to act as a back up source of energy for renewable energy sources of such as solar or wind. So even when the sun isn't shining or the wind isn't blowing, energy can still be provided. NuScale also suggests that their reactors could be used off the grid for places like military bases. Whatever role these reactors may play in a changing future, we will have to wait until the designs are approved to see what happens.

http://finance.yahoo.com/news/nuscale-submits-first-ever-small-140000073.html
Brumfiel, Geoff. "Miniaturized Nuclear Power Plant? U.S. Reviewing Proposed Design." NPR. NPR, 13 Jan. 2017. Web. 05 Feb. 2017.
http://www.npr.org/sections/the twoway/2017/01/13/509673094/miniaturized-nuclear-power-plant-u-s-reviewing-proposed-design
"NuScale Submits First Ever Small Modular Reactor Design Certification Application (DCA)." Yahoo! News. Yahoo!, 12 Jan. 2017. Web. 05 Feb. 2017.

MIT Nuclear Energy Lab Shut Down After Setting Records

by Mary-Catherine Riley

There is a joke amongst scientists that practical fusion power plants are just 30 years away–and always will be. However, in October 2016, Alcator C-Mod set a record for plasma pressure on its last day of operation at MIT's Plasma Science and Fusion Center. Other fusion experiments have reached the same temperatures, but none have reached the new record of two atmospheres.

Pressure is essential to nuclear fusion as each increased unit of pressure correlates to a squared increase in power. However, such high levels of pressure are hard to attain. Particularly because fusion chambers are complicated to create and maintain. The chambers must be over 50 million degrees, stable under intense pressure, and contained in a fixed volume.

ITER, a tokamak currently under progress in France, will operate at a lower magnetic field, but will be approximately 800 times larger in volume than the Alcator C-Mod. This €1 billion machine is expected to reach 2.6 atmospheres when it is in full operation by 2032.

Furthermore, a research lab in Germany is preparing to switch to a fusion device called a stellador, another alternative to the tokamak design. The Wendelstein 7-X (W7-X) is a 16-meter-wide ring ensconced in a maze of metals, coils, and wires. Once started, stellarators are easier to maintain and will not incur as many problems with metal bending under the pressure. However, they are infamously challenging to build. Lead of the German effort, Thomas Klinger states, "No one imagined what it means to build one." The complexity leads to an increased likelihood in mistakes and bugs, so we shall see if they meet their deadline.

Given the complexity and expense of building nuclear fusion machines, another large concern is the decrease in funding. In 2012, the MIT Research Lab ceased funding to Alcator due to budget constrictions. Following that decision, the U.S. funded the Alcator C-Mod for a three-year period, which ended September 30. It will be interesting to see how the nuclear sector changes in the coming years with projects in Europe expanding while projects in the United States are being cut.

(http://www.sciencemag.org/news/2015/10/bizarre-reactor-might-save-nuclear-fusion).
(http://news.mit.edu/2016/alcator-c-mod-tokamak-nuclear-fusion-world-record-1014).
Plasma Science and Fusion Center. "New Record for Fusion." MIT News. MIT News Office, 14 Oct. 2016. Web. 31 Mar. 2017.
Clery, Daniel. "The Bizarre Reactor That Might save Nuclear Fusion." The Bizarre Reactor That Might save Nuclear Fusion. American Association for the Advancement of Science, 12 Jan. 2016. Web. 31 Mar. 2017.
Chandler, David L. "A Small, Modular, Efficient Fusion Plant." MIT News. MIT News Office, 10 Aug. 2015. Web. 31 Mar. 2017.

Nuclear Plant Crack Detector

by Justin Wenig

Engineers from Purdue University have developed an automated system to detect cracks in the steel infrastructure of nuclear power plants. Aptly named CRAQ, the system can distinguish between cracks and other marks such as scratches. By helping nuclear engineers find cracks that otherwise would go unnoticed, the technology could help prevent nuclear leaks and hazardous incidents.

CRAQ uses recent advances in machine learning technology to identify cracks. First, the system takes multiple video clips of the steel surface on the inside of a nuclear reactor. Then, it processes the video using a statistical technique called Bayesian data fusion to assign confidence intervals to the likelihood that a change in the steel texture is a crack. Last, through the use of machine learning technology, the system filters out falsely detected cracks and predicts real cracks. While previous automated crack detection systems were unable to distinguish between cracks and other crack-like marks, CRAQ is able to distinguish between the marks by focusing on the texture-gradient of the image.

The system could reduce the cost of nuclear power plant maintenance. To illustrate, a 1996 incident at the Millstone nuclear power plant in Connecticut caused by a crack cost $254 million. The system could also decrease the maintenance cost of detecting cracks. According to research from Purdue, current inspection practices led by a manual operator are time-consuming and expensive. By automating the crack detection process to CRAQ, nuclear power companies could save a lot of money.

Interestingly, the engineers have also proposed that the system could be used to investigate the state of America's infrastructure. With a host of infrastructure initiatives that have recently been proposed in Congress, the system could help pinpoint which structures need repairs and which structures should be rebuilt.

http://onlinelibrary.wiley.com/doi/10.1111/mice.12256/full
https://www.sciencedaily.com/releases/2017/02/170217160934.htm

ENERGY POLICY AND ECONOMICS

Energy Policy

Clean Power Plan Faces an Uncertain Future
by Emily Audet

The EPA's Clean Power Plan (CPP), an enforcement plan of the Clean Air Act, establishes caps to carbon dioxide emissions of current power plants. The CPP has been controversial since its beginning. In December 2016, Texas and West Virginia led 24 states in urging President Trump to overturn the CPP. In response, many states and cities requested the preservation of the plan.

The implementation of the CPP is on hold as of January 2017 by order of the Supreme Court as the D.C. Circuit Court of Appeals.

On January 13, 2017, the EPA denied the majority of petitions challenging the plan or asking for a suspension of the plan's implementation. The EPA claims that many of the petitions rejected by the EPA on January 13th raised similar issues to petitions included in the comment period of the CPP's proposal. Of the 38 petitions asking for revisions of the plan, the EPA retained only 7 for further review. All 22 of the petitions advocating for a suspension of the CPP were rejected by the EPA on the grounds that the Supreme Court's stay of the plan already achieves this end.

As of January 2017, the CPP's future grows even more murky as it gets swept up in the uncertainty around environmental regulations in the new Trump administration. Throughout his campaign, Trump claimed that he would overturn the CPP, and an executive order from Trump could end the CPP, even before the courts release an official ruling on the plan. Scott Pruitt, the head of the EPA under the new Trump administration, has expressed his opinion that the EPA's strategy to lower carbon emissions should focus on individual technology innovations in firms to decrease emissions, rather than encouraging firms to move from coal to other, cleaner forms of energy, which the CPP currently does. If the new administration tried to weaken the CPP, environmental non-profits would likely bring the plan to court. If the CPP is overturned, the EPA continues to hold the authority to implement the Clean Air Act through other vehicles. As of January 2017, the EPA offers resources and other forms of support for states to implement similar regulations on the state-level.

[https://www.epa.gov/cleanpowerplan/clean-power-plan-existing-power-plants].
[http://voices.nationalgeographic.com/2017/01/05/fate-of-the-clean-power-plan-remains-uncertain/].

[https://www.law360.com/articles/880858/epa-denies-bids-to-reconsider-or-stay-clean-power-plan].

Rodriguez, Juan Carlos. "EPA Denies Bids To Reconsider Or Stay Clean Power Plan." Law 360. N.p., 13 Jan. 2017. Web. 24 Jan. 2017. https://www.law360.com/articles/880858/epa-denies-bids-to-reconsider-or-stay-clean-power-plan

Profeta, Tim. "Fate of the Clean Power Plan Remains Uncertain." National Geographic. N.p., 5 Jan. 2017. Web. 24 Jan. 2017. http://voices.nationalgeographic.com/2017/01/05/fate-of-the-clean-power-plan-remains-uncertain/

Holden, Emily. "What Could Replace the Clean Power Plan?" Scientific American. N.p., 23 Jan. 2017. Web. 24 Jan. 2017. https://www.scientificamerican.com/article/what-could-replace-the-clean-power-plan/

US EPA. "Clean Power Plan for Existing Power Plants." EPA. N.p., 12 Jan. 2017. Web. 24 Jan. 2017. https://www.epa.gov/cleanpowerplan/clean-power-plan-existing-power-plants

State-level Renewable Energy Regulations

by Emily Audet

States have often passed environmental regulations that extend past and are more stringent than federal regulations. With the current administration and Congress appearing to not prioritize sustainability nor clean energy regulations and legislation, pushes at state-level policy could be a viable political strategy for those concerned with advancing clean energy. As of January 2017, 29 states and Washington, D.C. have passed a renewable portfolio standard (RPS), a type of regulation that bolsters use and production of renewable energy. State-level RPSs significantly impact the nation's energy landscape—RPSs caused the creation of the majority of all renewable energy projects established from 2000 to 2017, and if states fully implement existing RPSs, a projected 40% of the energy for the whole country will come from renewable sources by 2050.

CleanTechnica published an article from a guest contributor on January 30, 2017 praising California, Iowa, Hawaii, New York, Minnesota, and Oregon for their strong clean energy policies The policies highlighted included RPSs, carbon emission reductions, investments into clean energy projects, and the phasing out of coal-fired power plants.

The article highlighted Hawaii's, Oregon's, and Minnesota's use of policy to transfer to renewable energy. Hawaii plans to use entirely renewable energy by 2045 and has mid-term goals established to help the state reach this ambitious, long-term goal. The Oregon government created a policy requiring utilities to transfer to 50% renewable energy by 2040. Minnesota's use of renewable energy has increased from 5% in 2005 to 21% of the state's power in 2015. The article also focused on California's and New York City's efforts to lower carbon emissions, as both bodies established regulations to cut carbon emissions by 40% by 2030.

Many of the states' legislation focuses on decreasing coal-fired energy. Oregon is the first state in the U.S. to create regulations with the aim to phase out all coal energy. Oregon plans to have moved completely away from coal-fired electricity by 2030. While enacted later, New York's regulations are more intensive, as the state intends to close all coal-based power plants by 2020. Coal remains a large part of Iowa's energy supply; however, in 2016, the state passed the development of the largest wind energy project in the country's history.

A 2016 report from the National Renewable Energy Laboratory and Lawrence Berkeley National Laboratory asserts that switching to renewable energy will save states in the long-run. The report is based on a study that ran a cost-benefit analysis of both existing RPSs and projected more expensive RPSs into 2050. Existing RPSs will cost states about $31 billion collectively by 2050; however, the study claims that these costs will be outweighed by the monetary gains of the environmental and health benefits spurred by decreased greenhouse gas emissions, pollutants, and water usage under existing RPSs. This study adds support to claims that RPS policies for states are economically, as well as environmentally, preferable.

"6 States Leading The Way In Clean Energy & Climate Legislation." CleanTechnica. N.p., 30 Jan. 2017. Web. 31 Jan. 2017. https://cleantechnica.com/2017/01/30/6-states-leading-way-clean-energy-climate-legislation/?utm_source=feedburner&utm_medium=feed&utm_campaign=Feed%3A+IM-cleantechnica+%28CleanTechnica%29
[http://www.nrel.gov/docs/fy17osti/67455.pdf].
[https://cleantechnica.com/2017/01/30/6-states-leading-way-clean-energy-climate-legislation/?utm_source=feedburner&utm_medium=feed&utm_campaign=Feed%3A+IM-cleantechnica+%28CleanTechnica%29].
[http://midwestenergynews.com/2017/01/09/report-benefits-of-state-renewable-energy-policies-far-outweigh-costs/].
Ferris, David. "Report: Benefits of State Renewable Energy Policies Far Outweigh Costs." Midwest Energy News. N.p., 2017. Web. 31 Jan. 2017. http://midwestenergynews.com/2017/01/09/report-benefits-of-state-renewable-energy-policies-far-outweigh-costs/
Mai, Trieu, Ryan Wiser, Galen Barbose, Lori Bird, Jenny Heeter, David Keyser, Venkat Krishnan, Jordan Macknick, and Dev Millstein. 2016. A Prospective Analysis of the Costs, Benefits, and Impacts of U.S. Renewable Portfolio Standards. NREL/TP-6A20-67455/LBNL- 1006962. Golden, CO and Berkeley, CA: National Renewable Energy Laboratory and Lawrence Berkeley National Laboratory. http://www.nrel.gov/docs/fy17osti/67455.pdf

Property Assessed Clean Energy: A Financial Tool to Encourage Sustainable Retrofitting
by Emily Audet

Buildings use up about half of the energy and three-quarters of the electricity used in the U.S. Per capita energy consumption in the U.S. is sizably large compared to the rest of the world, adding support to the necessity for energy retrofits to buildings in the U.S. Even though energy retrofits can save homeowners money in the long-term, the upfront costs for these upgrades can be so large as to discourage homeowners from investing in them. Property Assessed Clean Energy (PACE) is a financial tool that encourages homeowners to sustainably retrofit their homes by removing the upfront costs and allowing homeowners to pay back the price of these retrofits gradually, as their financial returns are realized. PACE pays for the cost of these upgrades upfront and the cost is added to the property's tax bill. In this way, property-owners can take up to twenty years to repay the price of these upgrades.

PACE applies to a broad range of properties, from homes to agricultural lands, and its funded energy improvements include solar panels, insulation, lighting, and many more options. If a relevant property is sold, the PACE credits remain with the property. PACE requires a public-private partnership to be implemented, as the repayment plan proceeds through taxation. First, states need to pass legislation allowing PACE, and then local governments decide whether to allow the program.

As of February 2017, 33 states have laws approving PACE. So far, PACE has funded $3.3 billion worth of energy retrofits and has produced over 29,000 jobs nationally. The Chief Sustainability Officer for San Diego, Cody Hooven, asserts that PACE retrofitting in private homes has significantly contributed to the realization of the city's Climate Action Plan, as over 2,000 San Diego homes have used PACE to install solar panels. Hooven says that PACE has been an important contributor to the city's 16% drop in electricity use in homes between 2010 and 2015.

[http://www.renewableenergyworld.com/ugc/articles/2017/02/23/pace-financing-is-the-solution-to-americas-silent-housing.html].

[http://americancityandcounty.com/blog/property-assessed-clean-energy].

[http://pacenation.us/what-is-pace/].

DeVries, Cisco. "PACE Financing Is the Solution to America's Silent Housing Crisis." Renewable Energy World. N.p., 24 Feb. 2017. Web. 28 Feb. 2017. http://www.renewableenergyworld.com/ugc/articles/2017/02/23/pace-financing-is-the-solution-to-americas-silent-housing.html

Hooven, Cody. "Property Assessed Clean Energy." American City & County. Viewpoints, 22 Feb. 2017. Web. 28 Feb. 2017. http://americancityandcounty.com/blog/property-assessed-clean-energy

"What Is PACE?" PACENation. N.p., n.d. Web. 28 Feb. 2017. http://pacenation.us/what-is-pace/

Executive Order Rolls Back Many Climate Change Regulations

by Emily Audet

On March 28, 2017, Trump signed an executive order to dismantle many climate change policies that the Obama administration established in accordance with their participation in the international agreement formed at COP21 in Paris in 2015. These Obama-era policies targeted by the Trump administration include a recommendation for federal agencies and the military to include climate change when conducting environmental reviews, as well as an estimate of the holistic costs of carbon emissions to U.S. society. The Interior Department will also reopen the sale of federal land to coal mining, ending a moratorium on selling federal land for this use. Other parts of the executive order include weakening regulations governing hydraulic fracturing on federal lands, reviewing current policies surrounding methane emissions from oil and gas well sites, and repealing Obama's executive order for the federal government to mitigate climate change effects.

In addition to these more specific directives, the executive order includes broad language for federal agencies to find and name any policies that hinder job creation and economic growth. This broad directive could empower agencies to strip back more climate protections than are included in the original executive order under the name of economic development. The Trump administration uses the prioritization of economic growth and U.S. energy independence to justify its rollback of environmental protections. This executive order also pleases Trump's supporters that were formerly employed by the coal industry, as throughout his campaign, Trump promised to increase the demand for labor in the coal industry. A White House official spoke in support of this executive order by arguing that the federal government can balance job creation and environmental protection if the EPA focuses solely on protecting air and water quality, rather than on climate change.

While the Trump administration has not officially repealed its participation in the pledges taken at COP21 in Paris in 2015, the administration has been repealing many policies related to meeting these pledges. While campaigning, Trump promised that the U.S. would exit the Paris agreement; however, a senior White House official has said that the administration has not yet reached a decision as whether to stay or leave this agreement.

[http://www.renewableenergyworld.com/articles/2017/03/trump-to-cancel-obama-s-policies-aimed-at-paris-climate-pledge.html; http://www.cnn.com/2017/03/27/politics/trump-climate-change-executive-order/].

Dlouhy, Jennifer A. "Trump to Cancel Obama's Policies Aimed at Paris Climate Pledge." Renewable Energy World. N.p., 28 Mar. 2017. Web. 28 Mar. 2017. http://www.renewableenergyworld.com/articles/2017/03/trump-to-cancel-obama-s-policies-aimed-at-paris-climate-pledge.html

Merica, Dan. "In Executive Order, Trump to Dramatically Change US Approach to Climate Change." CNN. N.p., 28 Mar. 2017. Web. 28 Mar. 2017. http://www.cnn.com/2017/03/27/politics/trump-climate-change-executive-order/

California's Cap-and-trade Program Upheld in Court of Appeals

by Emily Audet

Under AB 32, California's climate change law, the state must lower its greenhouse gas emissions to 1990 levels by 2020. California's main implementation mechanism to reach this goal is a cap-and-trade program, which auctions off permits to firms to emit a certain amount of carbon, which firms can then trade amongst each other. If firms do not comply with the emissions levels established under the program, the state fines them or enacts other penalties.

California has been running its carbon permit auction since 2012, and has raised $4.4 billion in funds for the state since then. Revenue from cap-and-trade goes towards carbon emissions-reducing projects, including construction of a high-speed rail line for the state. Revenue in the year leading up to April 2017, however, has been low, only around $500 million, because debate over the program's legality has hurt its efficacy.

In early April 2017, the Third District Court of Appeal in Sacramento voted to uphold California's ability to implement its cap-and-trade program until at least 2020. The lawsuit pitted the California Air Resources Board (CARB), which implements the cap-and-trade program, against the California Chamber of Commerce and Morning Star, a tomato-processing firm. The Chamber of Commerce and Morning Star argued that the cap-and-trade program is a tax, which would make it unlawful since Proposition 13 requires taxes to be passed with a two-thirds majority vote, whereas AB 32 passed with only a simple majority. CARB claimed that implementing the program was within the agency's jurisdiction since it was not a tax. The Court sided with CARB and elaborated that no one has a right to pollute. This court decision is a large win for the state, and many environmental groups celebrated the win as well. The plaintiffs, however, could still appeal this decision to the California Supreme Court.

[https://www.scientificamerican.com/article/court-upholds-californias-cap-and-trade-program/].

[http://www.renewableenergyworld.com/articles/2017/04/california-court-of-appeals-upholds-california-s-cap-and-trade-program.html].

[https://www.arb.ca.gov/cc/ab32/ab32.htm].

California Environmental Protection Agency Air Resources Board. "Assembly Bill 32 - California Global Warming Solutions Act." CA.gov. N.p., 5 Aug. 2014. Web. 11 Apr. 2017. https://www.arb.ca.gov/cc/ab32/ab32.htm

Kahn, Debra. "Court Upholds California's Cap-and-Trade Program." Scientific American. N.p., 7 Apr. 2017. Web. 11 Apr. 2017. https://www.scientificamerican.com/article/court-upholds-californias-cap-and-trade-program/

Smith, Allison, and Parissa Florez. "California Court of Appeals Upholds California's Cap-and-Trade Program." Renewable Energy World. N.p., 10 Apr. 2017. Web. 11 Apr. 2017. http://www.renewableenergyworld.com/articles/2017/04/california-court-of-appeals-upholds-california-s-cap-and-trade-program.html

Global Tracking Framework Report Shows Little Progress on Sustainable Energy for All's 2030 Goals

by Emily Audet

In April 2017, the World Bank and the International Energy Agency (IEA), as the primary members of the Sustainable Energy for All (SE4All) Knowledge Hub, released the third edition of SE4All's Global Tracking Framework (GTF) report. This edition of the report investigated progress from 2012 to 2014 on three of SE4All's energy goals, intended to be met by 2030: (1) comprehensive, global access to electricity and clean cooking, (2) doubling the global energy efficiency improvement rate, and (3) doubling the proportion of renewable energy in global energy consumption. The report revealed poor progress in all fields except energy efficiency Riccardo Puliti, Senior Director and Head of Energy and Extractives at the World Bank, and Dr. Fatih Birol, IEA Executive Director, spoke of this report's findings as call-to-action for increased commitment to achieving these goals.

In 2014, 85.3% of the world's population had access to electricity, and 57.4% had access to clean cooking fuel and appliances. The GTF report found that the global rate of electrification is decreasing. At this rate, only about 92% of the world's population will have access to electricity by 2030. Kenya, Malawi, Sudan, Uganda, Zambia, and Rwanda were among the nations with the greatest increases in electrification, improving their access by 2 to 3 percentage points per year since 2012. The report found that some countries, including Afghanistan and Cambodia, rely heavily on off-grid solar energy to progress in their electrification efforts, demonstrating potential for renewable energy to be used in improving electricity access. The percentage of the global population using traditional biofuels in cooking has remained relatively stagnant since 2012. While global access to clean cooking has been lagging, individual countries, including Indonesia, Sudan, and Vietnam, have been experiencing significant improvements.

The GTF report found that renewable energy comprised 18.3% of all final energy consumption, which is far from their 2030 goal. SE4All will likely focus its future efforts on the heat and transportation industries, since these industries comprise a large portion of energy usage.

[http://www.iea.org/newsroom/news/2017/april/uneven-progress-on-achieving-access-to-sustainable-energy-for-all.html;
http://www.worldbank.org/en/news/press-release/2017/04/03/more-action-needed-to-meet-energy-goals-by-2030-new-report-finds].
World Bank. "More Action Needed to Meet Energy Goals by 2030, New Report Finds." Text/HTML. The World Bank. N.p., 3 Apr. 2017. Web. 18 Apr. 2017. http://www.worldbank.org/en/news/press-release/2017/04/03/more-action-needed-to-meet-energy-goals-by-2030-new-report-finds
"Uneven Progress on Achieving Access to Sustainable Energy for All." International Energy Agency. N.p., 3 Apr. 2017. Web. 18 Apr. 2017. http://www.iea.org/newsroom/news/2017/april/uneven-progress-on-achieving-access-to-sustainable-energy-for-all.html

City of Portland and Multnomah County Announce 100% Renewable Energy Plan

by Emily Audet

In early April 2017, the City of Portland and Multnomah County, which houses the city, shared their plan to consume only clean energy by 2050. These government bodies intend to convert all electricity to renewable sources by 2035 and then transition all other energy consumption to renewables by 2050. Deborah Kafoury, the Chair of Multnomah County, praised the goal because of its potential to improve the Portland's climate and environment, and to spur economic development. Portland's Mayor, Ted Wheeler, emphasized the need for partnerships between the public sector, private sector, communities, and non-profit and community organizations in order for this goal to be achieved.the City of Portland stated that it will encourage climate- and environmentally-friendly industry within the region, continue implementation of the Climate Action Plan, and withstand weak federal carbon regulations in order to realize the goal.

The announcement comes at a time when several U.S. cities are stepping up and committing to renewable energy while federal carbon regulations regress in strictness. As of April 2017, Portland joins 25 other cities and 90 large companies in its goal of reaching entirely renewable energy. Just before Portland's announcement, Chicago's Mayor shared his plan to power city-owned buildings entirely with renewable energy by 2025. Portland's mission, however, stands apart from many other cities' plans in its inclusion of public transportation, as the city aims to use only electric vehicles by 2050.

This new plan is part of a legacy of climate activism in the Portland government. Portland was the first city in the country to develop a plan to decrease carbon emissions in 1993, and in 2017, Portland's carbon emissions are 21% less than they were at 1990. Portland currently uses renewable energy to power all of its city buildings. Multnomah County plans to acquire renewable energy to cover its facilities, starting in 2018.

[http://www.utilitydive.com/news/portland-commits-to-100-economywide-renewable-energy-by-2050/440335/]. [http://www.cnbc.com/2017/04/12/portland-commits-to-100-percent-renewable-energy-by-2050.html].

Frangoul, Anmar. "Portland Commits to 100 Percent Renewable Energy by 2050." CNBC. N.p., 12 Apr. 2017–12 Apr. 400. Web. 25 Apr. 2017. http://www.cnbc.com/2017/04/12/portland-commits-to-100-percent-renewable-energy-by-2050.html

Walton, Robert. "Portland Commits to 100% Economywide Renewable Energy by 2050." Utility Dive. N.p., 12 Apr. 2017. Web. 25 Apr. 2017. http://www.utilitydive.com/news/portland-commits-to-100-economywide-renewable-energy-by-2050/440335/

World Bank accused of incentivizing fossil fuel industries across the developing world

by Lauren Bollinger

The World Bank has been incentivizing fossil fuel dependence across the developing world, despite commitments to cut funding in such sectors, charges a January 2017 report by the advocacy group Bank Information Center (BIC).

The report, which examines the Bank's support of coal, gas, and oil projects in Peru, Indonesia, Egypt, and Mozambique, points out a contradiction between its pronouncements on climate change and its lending activities. The Bank has notably promised to work towards reducing subsidies for fossil fuels while incentivising investments in renewable energy. Most notably, in 2013, the Bank vowed to end virtually all support for the creation of coal-burning power plants, supporting them only in "rare circumstances" where there are no viable alternatives. Nonetheless, the BIC argues the World Bank has knowingly funded national policies to subsidize such fossil fuel industries.

The BIC's report comes after similar reports in October of last year by several US and Europe-based advocacy groups, on World Bank-backed coal projects throughout developing countries in Asia, from Bangladesh to the Philippines. In the Philippines, where the Bank has funded at least 20 new coal projects since 2013, such projects have drawn widespread criticism from human rights and indigenous advocacy groups, as the country's coal industry has resulted in an estimated thousand premature deaths annually and the displacement of thousands of indigenous peoples.

International finance institutions like the World Bank, which facilitate the loaning of millions of dollars to developing nations annually, carry immense political and economic clout in the developing world.

"World Bank accused of incentivizing investments in fossil fuels through $5B policy loans portfolio."
https://www.devex.com/news/world-bank-accused-of-incentivizing-investments-in-fossil-fuels-through-5b-policy-loans-portfolio-89528
"World Bank accused of funding Asia 'coal power boom'"
http://www.aljazeera.com/news/2016/10/world-bank-accused-funding-asia-coal-power-boom-161003045753947.html

Philippines makes moves to adopt Paris Climate Change Agreement

by Lauren Bollinger

Last month Philippines president Duterte signed the UN-drafted Paris Agreement after months of debate by his administration. The agreement, drafted during the 2015 United Nations Climate Change Conference, seeks to cut global dependence on fossil-fuels, with the goal of limit global warming to below 2 degrees Celsius. His decision marks the first step towards ratifying the plan, which if adopted, would hold the Philippines accountable to cut greenhouse-gas emissions by 70% by 2030.

This move comes after months of criticism of the plan, which Duterte contends unfairly holds developing countries such as the Philippines to the same standards as the largest polluters such as China and the United States, without the needed financial support. The Philippines, like many other developing countries, depends heavily on industries such as coal and lacks the proper infrastructure to meet such requirements.

The Philippines, after all, is a leader and former chair of the Climate Vulnerable Forum (CVF), an international organization comprising of countries that are disproportionately affected by the consequences of global warming. The twenty-two countries in the CVF, including over twenty countries such as Afghanistan, Ethiopia, and Vietnam, are among the first to be affected by climate change, and the last the recover from climate-change related environmental disasters. The CVF was formed to increase the accountability of industrialized nations for the consequences of global climate change, which there are widely believed to have caused.

Duterte's decision follows prior efforts by the country to push for the approval of the Paris Agreement and the inclusion of provisions beneficial to climate vulnerable countries.

Duterte set to sign PH consent to Paris Agreement
http://globalnation.inquirer.net/152230/duterte-set-sign-ph-consent-paris-agreement
Duterte to uphold Paris Agreement
http://globalnation.inquirer.net/148972/duterte-to-uphold-paris-agreement
Philippines to ratify Paris Climate Deal by July
http://www.rappler.com/nation/157808-philippines-to-ratify-paris-agreement-climate-
    change-july-2017

Trump to Sign Executive Order Reversing Obama-Era Climate Policies

by Lauren Bollinger

Trump is set to sign a sweeping executive order later Tuesday, March 28th, 2017, that would dramatically change the government's approach to climate change and environmental regulations. The plan represents a near complete reversal of several Obama-era climate policies. Specifically, the executive order is expected to roll back six of Obama's executive orders on climate change and regulating carbon emissions, from gutting the Clean Power Plan—which set a deadline of reducing carbon pollution from existing power plants by 32% by 2030 —to ending a moratorium on coal leases on public land, to rolling back methane restrictions.

Trump's decision follows through with longstanding campaign promises to curtail government involvement regarding environmental regulation.

The executive order also follows Trump's changes to the EPA, namely his selection of Scott Pruitt to EPA Administrator—who has a reputation for suing the agency over which he now presides—and his proposal to cut EPA funding by 31% in his proposed national budget.

As CNN reported on Monday, Trump's policy send a powerful message about his stance towards climate change and strengthening the U.S. coal industry and economy . "It is an issue that deserves attention," a White House official said of climate change. "But I think the President has been very clear that he is not going to pursue climate change policies that put the US economy at risk. It is very simple."

Whether Trump's decision will result in more coal jobs, only time will tell. As the Boston Globe reports, most analysis shows that such moves will only slow down the loss of jobs in the industry, not increase them, given the abundance of the cheaper alternative of natural gas.

In Executive Order, Trump to Dramatically Change US Approach to Climate Change
http://www.cnn.com/2017/03/27/politics/trump-climate-change-executive-order/
The Giant Trump Climate Order is Here
https://www.theatlantic.com/science/archive/2017/03/trump-climate-eo/520986/
Trump to sign order on Tues easing energy regulations: Officials
http://www.cnbc.com/2017/03/27/trump-to-sign-order-on-tuesday-easing-energy-regulations-officials.html

Pacific Islanders Fear for Future Amidst Trump's Climate Rhetoric

by Aurora Brachman

The future of Pacific Island nations requires the United States solidarity on climate change action to protect people in vulnerable developing nations from the environmental destruction they imminently face. As a result of rising sea levels and changing weather patterns, a consequence of global warming, Pacific Islands have begun to experience significant costal erosion and increasingly severe natural disasters that threaten the continued existence of these small island nations. Unfortunately, United States president Donald Trump has endorsed skepticism about human contributions to climate change and his climate policy as of January is consistent with these views. In Trump's 100-day action plan, which he issued during his campaign, he claims intentions to cancel billions of dollars in funds to the United Nation's climate change programs, which assist people in developing countries. He has also vowed to approve trillions of dollars' worth of energy projects involving shale, coal, natural gas and oil, all industries that perpetuate climate change and whose continued use pose a threat to the future of these island nations. Recently, many Pacific Island nations were the recipients of the United Nations Green Climate Fund (GCF), the largest climate fund that is committed to aiding developing countries in coping with climate change. The United States is the largest benefactor to the GCF, contributing 3.5 billion dollars to date; this could severely suffer under Trump's administration. One of the most vulnerable Pacific Islands, the small atoll nation of Tuvalu, expressed deep concern and dismay at Trump's rhetoric. Tafue Lusama, The General Secretary of the Tuvalu Christian Church and a vocal global advocate for climate action recently stated, "It is sad for us who rely on the United States to do the right thing and to hear the President embarking on the opposite path, which is ensuring our destruction,". In the coming months and years, we will see what climate policies President Trump has in store. If his policies continue to halt and erode action against climate change and endorse further reliance on fossil fuels, the fate of these Pacific Islands may be sealed.

Wilson, Catherina. "Pacific Islanders Call for U.S. Solidarity on Climate Change." Pacific Islanders Call for U.S. Solidarity on Climate Change | Inter Press Service. Inter Press Service News Agency , 19 Jan. 2017. Web. 23 Jan. 2017.
(http://www.ipsnews.net/2017/01/pacific-islanders-call-for-u-s-solidarity-on-climate-change/)

"Low-lying Micronesia Hopes Trump Reconsiders His Stance on Climate Change." VOA. Radio New Zeland, 18 Jan. 2017. Web. 23 Jan. 2017. (http://www.radionz.co.nz/international/pacific-news/322635/renewed-pleas-for-trump-to-change-climate-change-stance)

Brexit Could Derail EU Climate Goals

by Aurora Brachman

The European Union's attempts to fight climate change by reducing greenhouse gas emissions could be derailed by Brexit. This is because Britain's decision to leave the EU could disrupt the European Carbon Trading Scheme (ETS), a policy for greatly decreasing carbon emissions throughout Europe. Britain had previously pledged €2 billion for the project. Without this funding it is unlikely the project will be able to survive. The ETS would place a cap on how much carbon emissions a member state is allowed to produce. Companies that produce more than their allotted amount will have to pay for each additional ton of emissions they create. They can also buy and sell the amount of carbon they can produce, a system intended to encourage companies to reduce their emissions.

Because of Brexit there is now great fear that the plan could stop functioning all together. Much of the funding Britain was going to contribute would have helped Eastern European nations modernize their antiquated energy generators and manufacturing companies. This was a necessary first step for these nations to be able to fairly participate in ETS.

There is also fear that without Britain's financial support all of the EU's targets for combating climate change will become impossible to reach. Many fear the derailing of Europe's goals compounded with Donald Trump's anti-climate change policies could be disastrous for global efforts to make any serious combat against climate change.

Currently 10% of carbon emission allowances are going to be transferred to poorer countries within the EU. Because of its size, Britain would be contributing roughly 100 million allowances. Without Britain's contribution it is likely the lower income states could reject the climate targets set by the ETS.

Khan, Shehab. "Brexit could 'derail' EU attempts to fight climate change and reduce greenhouse gas emissions, say MEPs." The Independent. Independent Digital News and Media, 08 Feb. 2017. Web. 07 Mar. 2017. (http://www.independent.co.uk/environment/brexit-latest-news-derail-eu-attempts-climate-change-greenhouse-gas-emissions-meps-ets-a7569896.html)

Zamen, Peter. "The EU ETS and Brexit." Environmentalistonline.com. N.p., 14 Sept. 2016. Web. 07 Mar. 2017. (https://www.environmentalistonline.com/article/eu-ets-and-brexit)

Trump Unveils Budget Plan That Eliminates Programs to Combat and Study Climate Change

by Aurora Brachman

President Donald Trump's proposed federal budget is a heavy blow in the fight against climate change. The President's plan would end programs to cut domestic greenhouse gas emissions, eliminate diplomatic work to halt climate change, and stop the scientific study of the climate. Before unveiling this budget, the Trump administration had not taken a definitive stance on climate change. Trump has indicated that he believes climate change is a hoax but has also expressed a potential interest in taking steps to control climate change. Within the administration, Secretary of State Rex Tillerson has acknowledged the existence of climate change. However, Scott Pruitt the EPA Administer denies that $CO2$ is causing global warming.

The head of Trump's Office of Management and Budget, Mick Mulvaney, stated during a press briefing after the reveal of the proposed budget that "We're not spending money on that anymore" and "We consider that to be a waste of your money to go out and do that."

Within the EPA this budget would end funding to the Clean Power Plan, Obama's policy to reduce greenhouse gas emissions from power plants, as well as international climate change programs and climate change research and partnership programs. This budget also makes good on Trumps promise to end payments to the United Nations Green Climate Fund and its two previous Climate Investment Funds. The green climate fund is the U.N. attempt to help countries develop low-emission energy technologies and adapt to climate change. This proposed budget plan has been described as "eviscerating all of the programs that we had created over the course of the last eight years."

Thankfully, Congress will get the final say on the approval of this budget and many democratic congressmen have confidence in Congress to make the right decision. Rush Holt, a former new jersey congressmen and the CEO of the American Association for the Advancement of Science stated "Congress has a long bipartisan history of protecting research investments". He is hopeful that because of this the budget will not pass.

Greenfieldboyce, Nell. "Trump's Budget Slashes Climate Change Funding." NPR. Southern California Public Radio, 16 Mar. 2017. Web. 20 Mar. 2017
(http://www.npr.org/sections/thetwo-way/2017/03/16/520399205/trumps-budget-slashes-climate-change-funding)
Plumer, Brad. "Trump's budget envisions a US government that basically does nothing about climate change." Vox. Vox, 16 Mar. 2017. Web. 20 Mar. 2017
(http://www.vox.com/energy-and-environment/2017/3/16/14943826/trump-budget-proposal-climate-change)

Governments of Scotland and California Will Work Together to Combat Climate Change

by Aurora Brachman

The Scottish Government and the Government of California have signed an agreement to work together in the fight against climate change. Scotland's first minister, Nicola Sturgeon, and the governor of California, Edmond G Brown, met to discuss how the two administrations could work together. The governments signed a memorandum stating their commitment to the Under2 MoU, which both governments signed. This is an agreement to reduce greenhouse gas emissions by 80 to 95% below the levels they were at in 1990, or to cap emissions to under two metric tons per capita by the middle of the century.

The hope is that the two governments will be able to work together to help one another achieve the Under2 MoU goals. California has already demonstrated its commitment to meeting these energy goals. California aims to have 50% of its energy come from renewable sources by 2013. In the next 15 years the state also hopes to cut petroleum use in cars and trucks by 50%.

Both governments have an interest in greater investment in offshore wind energy and the two leaders discussed how their governments could share their knowledge and experience in developing and implementing this technology. California and Scotland have been world leaders in investing in renewable energy and they hope with their combined efforts they will be able to make impressive strides in continuing this practice. Despite the rhetoric of the Trump administration, this meeting signaled a continued commitment to fighting climate change both within the United States and abroad.

"Sturgeon signs climate agreement with California." BBC News. BBC, 03 Apr. 2017. Web. 04 Apr. 2017.
(http://www.bbc.com/news/uk-scotland-scotland-politics-39485807)
Frangoul, Anmar. "California and Scotland join forces to fight climate change." CNBC. CNBC, 04 Apr. 2017. Web. 04 Apr. 2017.
(http://www.cnbc.com/2017/04/04/california-and-scotland-join-forces-to-fight-climate-change.html)

New Report States We Must Halt Carbon Emissions by 2020 to Save Our Climate

by Aurora Brachman

The United Nations Framework Convention on Climate Change recently released a report entitled "2020" that creates a timeline for how we can still attain a safe and stable climate. In the report it is stated that we need to achieve certain goals by the year 2020, which will be when the earth will reach the fatal turning point in our climate. The report states that by 2020 carbon emissions in our atmosphere must cease to rise if we want to maintain a possibility of keeping global climate change below 1.5 degrees Celsius, the threshold established by the Paris Agreement.

The "2020" report states that we are currently on a trajectory to meet this goal. Largely thanks to China's economic transformation and the boom in renewable energy production globally, carbon emissions have plateaued. According to the report, to remain on this trajectory by 2020, renewable energy must beat out coal in all major energy markets. Deforestation must be quickly halted and the restoration of deforested land must have already begun. Every Fortune 500 company needs to be committed to meeting the goals of the Paris agreement, and capital markets need to have doubled their investments in zero emissions technologies.

All in all, "2020" is a report that confirms there is still a possibility for us to avoid the most devastating effects of climate change. However, we must do everything we can to ensure we meet the goals of the Paris Agreement by 2020. For it is in that year that the fate of our planets climate may be decided.

Mann, Michael E. "Climate Change: Trump Must Curb Carbon Emissions by 2020." Time. Time, 08 Apr. 2017. Web. 12 Apr. 2017.
(http://time.com/4731632/climate-change-2020-trump/)
"Climate Turning Point: 2020 identified as 'game-changing' year for global action." Edie.net. Edie Newsroom , 09 Apr. 2017. Web. 12 Apr. 2017.
(https://www.edie.net/news/9/Climate-Turning-Point--2020-identified-as--game-changing-
-year-for-global-action/)

Call for a Holistic Understanding of Energy Consumption in Urban Cities

by *Alejandra Chávez*

The article begins by explaining that 80 percent of the world's total energy production is consumed by urban areas, which are expanding and becoming increasingly complex. The largest energy-consuming areas are residential and commercial buildings, which are plentiful in urban areas and account for about one-third of the world's total energy consumption. Although energy efficiency initiatives and renewable energy investments are often common in residential and urban buildings —mainly for economic reasons— the authors stress that a "holistic" understanding of all the factors that influence consumption rates must be developed.

For example, the U.S. Energy Information Administration projected in their 2015 Annual Energy Outlook report that residential energy consumption will decrease by 0.3% and that commercial energy consumption will increase by 0.6%. These projections are not taking into consideration the location of the buildings, which, if spatial dependencies happen to be found, could cause completely different projections. Additionally, because the population of urban areas is expected to increase by nearly 70% by 2050, there will most likely be a rise in the number of human activities and developments that will significantly increase our energy consumption. This is especially true now that Exxon Mobile has estimated that the total energy demand of the world is expected to rise by more than 25% between 2010 and 2040.

The significance of urban spatial effects on building energy consumption was further explored by taking a closer look at the records of human mobility and energy consumption, over the course of a month, in Greater London and the City of Chicago. London and Chicago were chosen because of their urban location, population, and data availability. The article's results demonstrate that current energy consumption does not take into consideration "intra-urban human mobility," which along with spatial dependency, can significantly affect energy consumption predictions.

Mohammadi, Neda, John Taylor, and Yan Wang. "Towards Smarter Cities: Linking Human Mobility and Energy Use Fluctuations across Building Types." Proceedings of the 50th Hawaii International Conference on System Sciences. 2017.
ScholarSpace (http://hdl.handle.net/10125/41497)

Australian Trial Allows Neighborhoods to Monetize Excess Renewable Power

by Alejandra Chávez

A new government-backed initiative is officially to be one of Australia's first trials allowing homeowners and businesses to monetize solar and battery renewable energy and storage. The pilot program is called the Distributed Energy Exchange (deX) and officially launched in February of 2017 with the commitment to "change the way energy is produced, traded and consumed at a local level in Australia". Participants will be able to effectively 'rent out' or trade their excess electricity in a digital marketplace developed by deX. The marketplace will combine separate and small solar panels into one large virtual power plant, essentially also turning neighborhoods into power plants. A system of smart devices and software will enable excess solar and battery power to be routed to where it is most needed, making use of spare charges and automatically compensating homeowners accordingly.

The trial will be tested in two locations, encompassing a total of 10,000 households. Since around 1.6 million of the rooftops in Australia are already equipped with solar panels (one of the highest densities in the world, per capita) and battery storage is increasingly becoming more popular, innovative power distribution can help manage spikes in demand without building new structures. This is especially true because about 16 percent of the renewable electricity that is generated in Australia comes from solar panels, so the deX trial is taking advantage of a huge, untapped resource. It is estimated that the renewable energy contribution will increase between 20 and 50 percent of all electricity. The federal minister for the environment and energy, Josh Frydenberg, believes this project creates a "two-way interface between energy consumers and local network operators".

Similar 'virtual power plants' efforts have been made in the United States and Germany, but there has yet to be a wide-scale distribution of this kind of technology anywhere.

[https://www.theguardian.com/sustainablebusiness/2017/feb/23/ australian-consortium-launches-world-first-digital-energy-marketplace-for-rooftop-solar].

[https://www.theguardian.com/sustainable-business/2017/feb/23/australianconsortium-launches-world-first-digital-energy-marketplace-for-rooftop-solar]. Condliffe, Jamie. "A Renewables Marketplace Promises to Turn Neighborhoods Into Power Plants." MIT Technology Review. MIT Technology Review, 23 Feb. 2017. Web. 27 Feb. 2017.

MIT Technology Review (https://www.technologyreview.com/s/603718/a-renewables-marketplace-promises-to-turn-neighborhoods-into-power-plants/?set=603711)

Eldredge, Barbara. "In Australia, new program lets homeowners 'rent out' solar power and storage." Curbed. Vox Media, 24 Feb. 2017. Web. 27 Feb. 2017.

Curbed (http://www.curbed.com/2017/2/24/14720886/australia-renewable-energy-solar-power-dex)

Nogrady, Bianca. "Australian consortium launches world-first digital energy marketplace for rooftop solar." The Guardian. Guardian News and Media Limited, 22 Feb. 2017. Web. 27 Feb. 2017.

The Guardian (https://www.theguardian.com/sustainable-business/2017/feb/23/australian-consortium-launches-world-first-digital-energy-marketplace-for-rooftop-solar)

China in Transition-Searching for Sustainability

by Dominique Curtis

Coal is 66% of China's total energy consumption leading China to be one of the largest greenhouse gas emitters in the world. In a recent journal article Xiaoxia Zhou explains how China's growing economy, urbanization, and industrialization come at the steep cost of environmental pollution and the exhaustion of China's natural resources. Researchers, policy makers, scientists, and politics in China are scrambling to find a solution to these problems. They are making progress. China's energy development plan states that by 2020 their non-fossil energy will rise from it's current 9.8% to 15% (Guo, 2016). The question up for debate now is how? Outsiders suggest they should just go green and follow in the footsteps of other countries, but I think China is worried about a different kind of green. China is a coal-resourced based economy. They are focused on what clean energy will look like for their economy. The big question that Pibin Guo addresses in his article is "How does a coal resource-based economy transition to a sustainable economy?" Some researchers and policymakers have suggested clean coal technology and giving a positive connotation to coal. Guo says that China should accelerate their development of clean energy like natural gas, nuclear power, and renewable energy (Guo, 2016). Other suggestions have focused on policy and regulation changes. Many have also suggested the initiation and inclusion of more energy technology innovation. A few have suggested that China invest in clean coal technology. Something that all have agreed on is that the transition from a coal resource economy to a sustainable economy will also be a cultural and social transition. Zhou stated that it will require an environmental transition socially and politically. Policies and regulations for clean energy in China are on the way but this transition must also include individuals awareness of environmental quality and what Zhou calls a low carbon life.

Guo, Pibin, Ting Wang, Dan Li, and Xijun Zhou. "How Energy Technology Innovation Affects Transition of Coal Resource-based Economy in China." Energy Policy 92 (2016): 1-6. Web.

Zhou, Xiaoxiao, and Chao Feng. "The Impact of Environmental Regulation on Fossil Energy Consumption in China: Direct and Indirect Effects." The Impact of Environmental Regulation on Fossil Energy Consumption in China: Direct and Indirect Effects. Journal of Cleaner Production, 27 Oct. 2016.

Learning From the Students: Young Energy Innovators

by Dominique Curtis

Many university start ups across the midwest are competing for $50,000 from Clean Energy Trust. Clean Energy Trust helps launch, fund and grow Midwest clean energy companies to ensure a more prosperous, sustainable future for generations to come. Clean Energy Trust is a Chicago-based energy innovation non-profit that is sponsoring a Cleantech University Prize competition. The competition's goal is to support the next generation of energy innovators through mentorship and funding.

As of March 7,2017 when this was written the midwest competition was down to its eight finalists. These university students' ideas range from advanced batteries, drones, water purification, blockchain for energy, and many more. A University of Minnesota startup is working on developing advanced energy storage devices such as lithium air batteries. They think that technologies such as these will help speed up the transition to a electric economy. Michigan State focused on aircraft and chose to analyse their rotor blades. They came up with the idea to have bladeless aircraft and their test project is bladeless drones that use closed impeller vanes. These types of technology advances can lead to aircraft that can also be sub-aquatic. Case Western Reserve University found a safer, environmentally friendly and cost-efficient way to dispose of car tires.

Over the coming months these university start-ups will work with and receive mentorship from Clean Energy Trust experts. The students will get help on refining their ideas, innovations, and business models. The CEO of Clean Energy Trust stated that "The diverse cross-section of universities and technologies represented by the finalists demonstrate that the Midwest's universities continue to lead on using science and innovation to solve some of the biggest challenges we face" (Hustad, 2017). Many universities and companies are beginning to turn to the students and young innovators for answers and solutions to the world's energy crisis.Investing in the new generation will help preserve the world's resources for future generations.

"Clean Energy Trust Challenge." Clean Energy Trust. N.p., 2017. Web. 01 Mar. 2017.

Hustad, Karis. "These Student Startups Will Compete for $50K at Clean Energy Trust's Cleantech UP Competition." Chicago Inno. Streetwise, 05 Jan. 2017. Web. 07 Mar. 2017.

http://dx.doi.org/10.1371/journal.pone.0169045

Congressional Review Rolls Back Environmental Regulation

by Ethan Fukuto

On Wednesday, February 1, the Republican-majority Congress began its roll back of Obama-era environmental regulations through the Congressional Review Act (or the CRA). Enacted in 1996, the CRA allows for Congress to overturn regulations within a period of 60 session days, and is not subject to filibustering. The CRA's use in March of 2001 under the Bush administration, repealing a Clinton-era regulation on workplace injury, highlights perhaps the only timeframe when Congress can use the act successfully: the beginning of a new presidency.

This rollback comes under the guise of boosting jobs in the energy sector, with Republican officials asserting that these rules come at the expense of the job market. The 'Stream Protection Rule", intended to protect waterways from mining, was repealed in a vote of 228 to 194, allowing mining companies to dump waste into streams. The Rule specifically protected streams from the debris of mountaintop removal, a process in which mountaintops are scraped off to mine the minerals beneath. To date, more than 500 mountaintops in the Appalachia region have been removed, polluting the region and thereby increasing the risk for cancer and birth defects in nearby communities.

On Friday, February 3, the House repealed a rule on methane gas emission in a vote of 221 to 191, again with the intent of increasing jobs and overall energy production in the United States. The regulation on methane was predicted to decrease emissions in an amount comparable to moving one million vehicles off the road per year. Methane is 25 times more potent in trapping heat than carbon dioxide, and contributes around 9% of the greenhouse gas emissions in the U.S. Some have cited the regulation as unnecessary in an industry which already captures and sells emissions; the Western Energy Alliance reports that methane emissions have been reduced by 21% since 1990, those these numbers have been challenged by Representative Alan Lowenthal of California. Regardless of criticism, roll backs are finding little bureaucratic resistance in an administration keen to undoing the legacy of its predecessor.

Daly, Matthew. "House Overturns Obama Administration's Methane Gas Emission Rule". Time, 3 February 2017. http://time.com/4659832/congress-overturns-methane-gas-emission-rule/

Milman, Oliver. "Republicans target environmental rules protecting parks and limiting methane". The Guardian, 1 February 2017. https://www.theguardian.com/us-news/2017/feb/01/environment-republicans-congress-climate-methane-parks

Shapiro, Stuart. "The Congressional Review Act, rarely used and (almost always) unsuccessful". The Hill, 17 April 2015. http://thehill.com/blogs/pundits-blog/lawmaker-news/239189-the-congressional-review-act-rarely-used-and-almost-always.

Walsh, Deirdre. "GOP House votes to reject Obama administration stream protection rule". CNN, 2 February 2017. http://www.cnn.com/2017/02/01/politics/stream-protection-rule/

Repeal of Obama's Transparency Rule for Energy Companies

by Sagarika Gami

The Senate voted on Friday, February 3rd to nullify the "Disclosure of Payments by Resource Extraction Issuers" rule established by former President Obama's administration in 2010 and finalized by Securities and Exchange Commission on July 27, 2016. This rule was an anti-corruption effort. It was a part of the Dodd-Frank Wall Street Reform and Consumer Protection Act, which established new restrictions and safeguards post financial crisis and Recession. The rule required energy companies (oil, natural gas, coal, and mineral companies) to declare their royalties and government payments in their business dealings to the U.S. Securities and Exchange Commission. It was intended to be a transparency measure and method to deter corrupt business practices in resource-rich countries.

American oil and gas companies widely opposed the rule, thinking that they would be put at a disadvantage against private companies and companies not traded publicly in the US. GOP Senators also felt that the rule imposed unreasonable compliance costs, so Republican representatives Bill Huizenga of Michigan and Jim Inhofe of Oklahoma initiated the vote to repeal it. The Senate passed the repeal with a vote of 52 to 46, a vote split directly along party lines; it passed the House in a 235 to 187 vote. As of early February the resolution was awaiting approval from President Donald Trump who is expected to sign it.

Many critics call the resolution a part of the new administration's agenda to deregulate corporate America. Oxfam America, a non-profit working to end the injustice of poverty, released a statement by Isabel Munilla, a senior policy advisor for extractive industries at Oxfam, in which she condemned the passage of the resolution. She called out Republican senators to be voting for corruption under the guise of deregulation, expressing that, "In the five years after Section 1504 passed, more than $1.5 trillion should have been paid by oil companies to governments of some of the poorest on earth...but these payments will remain in the dark because of today's actions." Senator Elizabeth Warren also released commentary, calling the repeal a "giveaway" that will help "corrupt and oppressive foreign regimes and make it easy to funnel money to terrorists around the world."

Upi.com (http://www.upi.com/Top_News/US/2017/02/03/Senate-votes-to-kill-Obama-transparency-rule-for-energy-companies/3331486161147/)

Congress.gov (https://www.congress.gov/bill/115th-congress/house-joint-resolution/41)

Thestreet.com (https://www.thestreet.com/story/13977811/1/anti-corruption-rule-killer-for-energy-companies-heads-to-trump-s-desk.html)

oxfamamerica.org (https://www.oxfamamerica.org/press/oxfam-condemns-senates-shameful-gutting-of-bipartisan-anti-corruption-rule-section-1504/)

Trump's Initiative to Repeal Obama's Clean Power Plan

by Sagarika Gami

The Trump administration has prioritized repealing Obama's Clean Power Plan (CPP). Data from a new technology, Energy Policy Stimulator (EPS), shows that repealing CPP will increase costs to the U.S. economy, add more than a billion tons of greenhouse gases to the atmosphere, and cause more premature deaths as a result of inhaled particulate pollution. Specifically, repealing CPP will cause an increase of carbon dioxide equivalent emissions of 500 million metric tons in 2030 and 1200 million metric tons in 2050, potentially causing increased hurricanes, floods, and droughts. Accounting for increased capital, fuel, and operations and maintenance expenditures, net costs to the U.S. economy will exceed $100 billion by 2030 and almost $600 billion by 2050. The increased particulate emissions will cause more than 40,000 premature deaths in 2030 and more than 120,000 premature deaths in 2050.

EPS is an open-source computer model that was created to estimate the economic and emissions effects of different combinations of energy and environmental policies using non-partisan, published data from the U.S. Energy Information Administration (EIA), U.S. EPA, Argonne National Laboratory, U.S. Forest Service, and U.S. Bureau of Transportation Statistics. An interesting part of this technology is that EPS is freely available for public use.

The CPP is a set of rules created by the EPA to limit pollution from power plants by focusing on reducing carbon emissions. The aim is to reduce carbon dioxide emissions from existing power plants by 32% the 2005 levels. The policies work to reduce particulate pollution, which is responsible for thousands of heart attacks and respiratory diseases every year. The plan is the subject of a highly contested legal battle in the U.S. Court of Appeals for the District of Columbia Circuit to decide whether the plan's carbon-cutting requirements overstep constitutional boundaries.

Some proposed alternatives from the Trump administration are to either replace CPP with nothing so as to stop regulating carbon emissions from existing power plants at all or to rewrite CPP to be much weaker and scaled back. The second alternative would lower the carbon dioxide targets states are required to abide by.

Forbes.com (https://www.forbes.com/sites/energyinnovation/2017/02/23/clean-power-plan-repeal-would-cost-america-600-billion-cause-120000-premature-deaths/#2a9e51f33b78)

vox.com (http://www.vox.com/energy-and-environment/2017/2/23/14691438/trump-repeal-clean-power-plan)

washingtonpost.com (https://www.washingtonpost.com/news/energy-environment/wp/2016/11/11/trump-has-vowed-to-kill-the-clean-power-plan-heres-how-he-might-and-might-not-succeed/?utm_term=.64568d5310ea

Scott Weiner Proposes Tough Solar Panel Bill for California

by Genna Gores

With the inauguration of Donald Trump, and a cabinet full of climate deniers, it is now up to states to lead the charge in the fight against climate change. California, a trailblazer in climate change initiatives, will be very important in the upcoming four years to help lead the country on green initiatives. Even before the Trump Administration took office Scott Wiener, D-San Francisco—a newly elected California state senator—started proposing a new solar panel legislation.

This legislation bases itself off of a bill, *Better Roofs Ordinance*, that passed unanimously in San Francisco in 2016, which Wiener also proposed while he was while he was the County Supervisor. It required all new commercial and residential buildings between zero to ten stories to install solar panels on 15% of their roofs (taller buildings are excluded from the bill because powering them through solar panels poses too much of challenge to be required by city/state laws). With just this ordinance San Francisco is able to eliminate 26.3 million tons of carbon dioxide annually. San Francisco is one of four cities in the US that passed a solar panel bill, and if California passes this future legislation it will be the first state to take on this challenge.

Passing legislation in a small county like San Francisco is clearly much simpler than at the state level, especially a state as big as California. Currently, though, California already has legislation that requires all new residential and commercial buildings to have "solar ready" roofs, where Wiener's legislation takes it one step further by saying this buildings need to be "solar installed." California is already producing 15,000 mega watts of solar energy (enough energy to power 3.7 million homes), and enforcing solar panels can only increase renewable energy production.

It will also help California reach its goal of lowering greenhouse gas emissions by 40% below 1990 levels by 2030, which is a tough feat that needs harsh legislation like the *Better Roofs Ordinance*. As with any bill proposed there will be opposition, and the current argument is over cost of the solar panels. In California the housing market is skyrocketing, and people are worried that this ordinance will only raise prices. Builders will have to spend more money to install these panels, and thus the new cost will be put on buyers (with a higher selling price for buildings). Despite this opposition Wiener believes that the overall energy savings from having solar panels will outweigh the initial cost, and or builders can utilize third party companies that will own and maintain the panels.

[http://www.sfchronicle.com/bayarea/article/Scott-Wiener-models-state-solar-bill-on-S-F-law-10843577.php] [http://www.latimes.com/politics/la-pol-ca-jerry-brown-signs-climate-laws-20160908-snap-story.html]
[https://qz.com/665574/san-francisco-will-require-new-buildings-to-install-solar-panels/]
https://medium.com/@Scott_Wiener/lets-require-solar-panels-on-all-new-buildings-in-california-cb18fe9d9ec4#.b71kzqib6] [https://medium.com/@Scott_Wiener/lets-require-solar-panels-on-all-new-buildings-in-california-

cb18fe9d9ec4#.b71kzqib6] https://medium.com/@Scott_Wiener/lets-require-solar-panels-on-all-new-buildings-in-california-cb18fe9d9ec4#.b71kzqib6

http://www.latimes.com/politics/la-pol-ca-jerry-brown-signs-climate-laws-20160908-snap-story.html

http://www.sfchronicle.com/bayarea/article/Scott-Wiener-models-state-solar-bill-on-S-F-law-10843577.php

In Wake of Dakota Access Pipeline Controversy Seattle Moves to Break Ties with Wells Fargo

by Genna Gores

Over the past two decades Wells Fargo has been the primary financial service provider for, but with the current controversy over the Dakota Access Pipeline the city ~~Seattle~~ wants to end that partnership. Wells Fargo is an investor in the pipeline and the bank claims to have loaned $120 million to Energy Transfer Partners, one of the companies involved in constructing the pipeline. Critics, though, believe that this figure is too low and that Wells Fargo is giving more and providing other services that they are not willing to disclose.

The DAPL is a controversial crude oil pipeline that will cost $3.8 billion to construct. It will be a 1,170-mile pipeline running from North Dakota to Illinois. Most of the controversy stems from a specific section of the pipeline that goes through a section of the Standing Rock Sioux tribe's land that holds cultural and sacred significance. Furthermore, if an oil spill were to occur, it would be catastrophic for the tribe's water supply. The Tribe was not approached initially about this project even though it federal law mandates protection of their sacred land. For months protests occurred to reroute and or stop the pipeline and the Obama administration complied with this request. Now President Trump is ordering its completion.

As a way to fight back, a nine-member Seattle council unanimously passed legislation that would pull their $3 billion dollars out of Wells Fargo. While the Seattle community would not suffer the direct effects of the DAPL, the council feels it is important to join this fight. Seattle considers itself an activist and environmentally aware city and believes, "The example that we have set today can be a beacon of hope to activists all around the country seeking to change the economic calculus of corporations who think that investing in the Dakota Access pipeline will be good for their bottom line...We're making it bad for their bottom line." Even though this action most likely will not stop the construction of the pipeline, the Seattle city council sees the importance of fighting for the rights of the Sioux Tribe. It will end their relationship with Wells Fargo in 2018 when their current contract expires.

http://www.latimes.com/nation/la-na-seattle-divests-from-wells-fargo-20170206-story.html

[http://www.latimes.com/nation/la-na-seattle-divests-from-wells-fargo-20170206-story.html]

Budweiser Pledges New Renewable Energy Plan for 2025

by Genna Gores

Anheuser-Busch InBev, the large Belgian beer company that owns beers such as Stella Artois, Corona, and Budweiser, promises that all their purchased electricity will come from renewable energy sources by 2025. This change to renewables will lower the company's carbon footprint by 30%—this reduction is comparable to 500,000 cars being taken off the road. If this promise is upheld by the company then they will produce 6 terawatt-hours of clean energy, which is equivalent to 400 soccer fields worth of solar panels. The company plans to source 15%-25% of its energy from on-site technology and 75%—85% through direct power purchase agreements through other energy providers. These power purchase agreements are ultimately cheaper than grid electricity because they allow AB InBev to lock in their price on renewable energy, which protects them from the changing markets. AB InBev believes that most of their electricity will come from wind and solar energy, but this all depends on technology and which way the green-energy markets go.

By 2019, they will launch this new renewable program in their biggest brewery in Zacatecas, Mexico. The company signed a contract with Iberdrola, a renewable energy company, for 490 gigawatt-hours per year, and this partnership will give AB InBev all the purchased electricity they will need to run renewable production at this site. With this type of commitment, AB InBev, hopes to be the largest corporation to directly purchase renewable power. They hope this switch to renewables will save the company money on electricity costs, and demonstrate their commitment to being greener, not only for the natural environment, but to the people who drink their beer.

https://www.usatoday.com/story/money/nation-now/2017/03/30/your-budweiser-brewed-renewable-energy-2025/99815814/

http://www.anheuser-busch.com/newsroom/2017/03/anheuser-busch-inbev-commits-to-a-100--renewable-electricity-fut.html

A Greener Apple

by Siena Hacker

Apple Inc. released its 2017 Environmental Responsibility Report in time for Earth Day. The technology giant may be associated with iPods and Steve Job's turtlenecks, but the company actually prides itself on its commitment to sustainability. Lisa Jackson, the former administrator of the Environmental Protection Agency under President Obama, is now Vice President of Apple's Office of Environment, Policy, and Social Initiatives. Jackson is overseeing an incredibly ambitious goal: transitioning the company to only use recycled materials for their devices. For example, Apple is attempting to make the main logic board of the iPhone 6s completely from recycled tin solder. The company's closed-loop supply chain will draw support from Liam, Apple's line of disassembly robots, and Apple Renew, a program that encourages consumers to return their old products online or in stores for a gift card. The Apple Renew program then refurbishes and resells the device, or completely recycles the materials. In addition to their new recycled material programs, Apple has also been working to lower their carbon footprint. The average carbon emissions per product was 97 kilograms in 2016, compared to 137.2 kg in 2011. They have identified further ways to reduce the carbon footprint of their products. For example, their iPhone 7 uses 27% less virgin aluminum than the iPhone 6, which is notable because aluminum makes up almost 30% of Apple's manufacturing carbon footprint. The company has also been able to reduce their emissions by identifying ways to help their suppliers switch to renewable energy. They conducted 34 energy audits at their suppliers' warehouses in 2016, and suggested green switches that would save over $55 million annually. The report states that suppliers' switches to more efficient practices sequestered more than 150,000 metric tons of carbon dioxide equivalents in 2016. According to their report, Apple and their suppliers will produce and procure at least four gigawatts of new clean power by 2020, with two gigawatts of that being generated in China alone. Apple is setting a green example for other companies while also saving suppliers and themselves money. Hopefully other large companies will began to follow their lead!

2017 Environmental Responsibility Report:
     https://images.apple.com/environment/pdf/Apple_Environmental_Responsibilit
     y_Report_2017.pdf
Apple Renew: https://www.apple.com/recycling/
iLounge: http://www.ilounge.com/index.php/news/comments/apple-releases-2017-
     environmental-responsibility-report

India's Plan for Renewable Energy

by Cybele Kappos

India is currently dealing with many problems regarding electricity and pollution. In his 2017 article, Michael Lynch discusses how the poverty-stricken country's infrastructure is such that the population suffers from shortages of electricity and unreliable power supply. Moreover, despite government reports that 96–97% of villages have access to electricity, several millions are not actually connected to the grid. The country's electricity is defective to say the least. Nevertheless, India has set a target to be able to create 175 GW of renewable energy by 2021/2022. While this is a commendable commitment to address the problems of pollution due to coal-fueled plants and the inadequacy of the current situation of electricity, Lynch argues that this is the worst possible policy for a country like India. First, he argues that renewable energy is unpredictable and unreliable. This, he says, could aggravate the current problem of erratic connections to the grid and possibly increase the cost of power. Second, Lynch argues that India's electricity sector suffers from mismanagement and consequently, significant debt. The costs of renewable energy are site-dependent and he predicts that while there will be locations where renewable energy is economically efficient, renewable energy is typically more expensive than that from fossil fuels. Third, Lynch argues, other countries such as Germany and China have demonstrated how government mandate goals are ineffective and India cannot be expected to perform better. Lynch does not necessarily argue against the nation's initiative to reform energy, but suggests that it could either consider switching to natural gas or potentially modernize the existing coal-fueled plants, thereby reducing pollution. This, he argues, would both be economically more beneficial for a nation that is already suffering in this sector and also be met with less opposition by people whose land would potentially be taken in order to construct the new plants. Lynch concludes by saying that India should look to the failed examples of other countries to learn that this target will most likely fail and should switch to incentives to reduce emissions rather than create implausible goals.

Lynch, Michael. India's Renewable Investment Goal Seems Misguided. Jan 25, 2017
http://www.forbes.com/sites/michaellynch/2017/01/25/renewable-investment-
    announcement-from-india-seems-misguided/2/#4bd328b043f6
http://economictimes.indiatimes.com/news/politics-and-nation/indias-renewable-energy-
    targets-catch-the-attention-of-global-investors-still-need-ground-
    work/articleshow/53015707.cms

The Kingdom of Saudi Arabia's Vision 2030
by Dena Kleemeier

My family has worked for Saudi Aramco, a national petroleum and natural gas company based in Dhahran, Saudi Arabia for the past 16 years. In my time living in Saudi, I have experienced first hand the way in which the Kingdom uses/wastes their resources, providing oil to their citizens at a cheaper price than water, and subsidizing electricity to their populations. However with the decline in the price of crude (lost 67% of value since September 2014), the growing domestic oil demand of 7% per year, and the state of the environment, Saudi Arabia is in an awkward geopolitical situation, and is in need of comprehensive economic reform. Although I am critical, Dr. Mamdouh G. Salameh's article "Saudi Arabia's Vision 2030: A Reality or Mirage" was enlightening to read an article explaining reality of Crown Prince Mohammed Bin Salman's plan. The article reports three major points to the vision; 1.) Expand the non-oil sector of the economy (a good starting point, as over 85% of the KSA's revenues are from oil). 2.) Triple the share of non-oil exports to $100 bn by 2020, and 3.) Reduce unemployment, and create 3m jobs by 2020. However, in order to carry through this vision, the kingdom needs to diversify its economy, invest in food production, petrochemicals, solar energy, nuclear power, water desalination plants, and phasing out financial subsidies for gas, diesel and electricity. The specific strategies that the KSA intends on adopting are adding value to exported fossil fuels, introducing energy efficient measures, and withdrawing from the petrodollar. This success will depend on the influence and power that the prince possesses. From my experience seeing solar panels implemented at Saudi Aramco, dust has hugely influenced the effectiveness of the panels, to the point that the company doesn't want to pay to get the panels cleaned, thus leaving them unused. Vision 2030 is designed to create a 21st century Saudi economy virtually from scratch, which I don't see as currently being an economically viable option. It isn't clear which parts of Prince Mohammed Bin Salman's plan will be effective, and which parts will not be environmentally and economically viable.

Salameh, Mamdouh G., Saudi Arabia's Vision 2030: A Reality or Mirage (July 12, 2016). USAEE Working Paper, Forthcoming. Available at SSRN: https://ssrn.com/abstract=2808611

Past is Not Prologue

by Dena Kleemeier

The Governor of the Bank of England and Chairman of the Monetary Policy Committee, Mark Carney gave a speech at Lloyd's of London on the intersection of climate change and financial stability. His speech is from the point of view of the insurance industry, and of economists. He makes the claim that because climate change has the potential to affect insurance industries, these are the industries that must actively seek to tackle the issue. He moves on to outline climate change as a tragedy of the commons but also a tragedy of the horizon in that there will be costs to future generations that current generations are neglecting to address because they do not imminently affect them.

Carney outlines the paradoxical framework of climate change as an issue, and how early action will be less costly for the insurance industry in the long term. But, once climate change becomes a defining issue for financial stability, it will be too late. He used three specific, examples of channels by which climate change can and will affect financial stability: physical risks, liability risks, and transition risks. All three are preventable through early action. Throughout his speech, Carney's solution statements are broad and vague; "the more we invest with foresight, the less we must do in hindsight", and "we must develop frameworks to help the markets transition".

Ultimately the framework that he describes is creating a greater connection between scientists, policymakers, and firms so that the latter have enough scientific information to help climate policy be more like monetary policy in that economists look at trends in the market, and to encourage the funding of technologies that will help climate change. He concludes by saying, "by managing what gets measured, we can break this tragedy of the horizons", another vague and ambiguous statement.

"Breaking the tragedy of the horizon - climate change and financial stability - speech by Mark Carney." News and Publications. N.p., n.d. Web. 16 Apr. 2017.

New Proposal Seeks to Reduce the Numbers of People Driving to Work

by Kieran McVeigh

In recent years cities across the world have a tried a variety of policies to reduce the number of people using cars within the city limits, from citiBikes, to cleaning up public transportation, and even levying fines against citizens with the most polluting vehicles. In Washington DC, a new policy has been proposed to the city council, suggesting in effect, paying people not to drive to work. More specifically the proposed policy would require employers to provide equal incentives for people to not drive to work as they do for people driving to work.

The way this works is that many jobs offer free parking as a way to make their employees lives a little easier, this "free" parking actually serves as a powerful incentive for people to drive to work, but although "free" to employees, companies have to pay the costs of this parking, the new policy would require that any company offering free parking to its employees must provide a similar incentive to employees should they choose to commute to work a different way. Proponents argue that at its core this policy is about fairness, suggesting that when employers offer the benefit of free parking it unfairly subsidizes employees who drive to work, often those with more means. Although this line of argument has merit, another possible large effect of the policy is to incentivize cultural change that will lead to a more sustainable society, through reducing the amount we drive to work.

Research suggests that these types of programs can have a substantial impact. Studies in Los Angeles found a similar policy reduced the number of people driving to work alone by up to 13%. DC offers and ideal testing ground for proposals like these as it has a large subway system, a relatively small footprint as a city, local carpooling programs, and an abundance of Citibikes. If the policy is effective in DC, it could be adopted nationwide and start creating a large change in the amount of emissions and pollution nationwide.

Marshall, Aarian. "Want Commuters to Ditch Driving? Try Giving Them Cash Money." Wired. Conde Nast, 26 Mar. 2017.   Web. 27 Mar. 2017. https://www.wired.com/2017/03/want- commuters-ditch-driving-try-giving-cash-money/

Lazo, Luz. "D.C. wants employers to pay workers not to drive to work." The Washington Post. WP Company, 17 Mar. 2017. Web. 27 Mar. 2017. <https://www.washingtonpost.com/news/dr-gridlock/wp/2017/03/17/d-c-wants-employers-to-pay-workers-not-to-drive-to-work/?utm_term=.20cb9f515b20>.

How Puerto Rico's Energy Sector Can Revitalize the Island's Struggling Economy.

by Byron R. Núñez

The Commonwealth of Puerto Rico has more than $70 billion of debt, most of which can be attributed to the United States' decision to cut corporate tax breaks. The current financial crisis has created a mass exodus by U.S. companies and people from the Island. To ameliorate the situation, President Barack Obama signed the Puerto Rico Oversight Management and Economic Stability Act (PROMESA), which led to the creation of a committee design to manage the island's finances. This economic instability has forced Puerto Rico's energy sector to reinvent itself and become more cost-effective and efficient.

Currently, Puerto Ricans pay two to three times more for electricity than average Americans. The strongest factor for the island's high energy costs is that 80% of the energy used on the island comes from imported petroleum as the island itself does not produce nor refine crude oil. Sustainable energy is key to Puerto Rico's future as the island hopes to comply with a Renewable Energy Portfolio Standard (REPS) that hopes to supply 20% of electricity with green energy by 2035. One company that is hoping to revitalize the island's struggling economy though the energy sector is Green Kinetic Power (GKP), LLC.

GKP, LLC is a Puerto Rican based company that created the Traffic Energy Bar System (TEBS). TEBS is a patented technology that captures kinetic energy from the weight of moving vehicular traffic and converts it into electrical energy. The electricity is produced by electromagnetic generators that can be interconnected and synchronized in a scalable manner to generate large-scale clean energy. The mechanism is installed in the roads and is activated when a vehicle travels over the unit. This Puerto Rican company is hoping to reduce carbon generation by providing clean energy.

One clear advantage of this technology is that it can also provide traffic data (frequency of vehicles, their weight, and their speed) for public authorities to use. Puerto Rico's economy needs major reforms for the island to get rid of its debt and solve its energy crisis. TEBS hopes to do its part to produce clean energy while also creating green jobs and contributing to the islands economic development and growth. This technology is now also available in Europe and the United States.

Uncova
http://uncova.com/puerto-rico-turns-to-tech-and-entrepreneurialism-to-revitalize-the-economy

How the DRC Hopes to Become a Global Power House

by Byron R. Núñez

Lack of access to consistent supply of electricity has negatively impacted the growth of many developing countries. In the Democratic Republic of Congo (DRC), access to electricity remains at around 15%, which is relatively low even when compared to the pan-African standard of 40%. Through numerous projects, such as the Great Inga dam, the DRC hopes to increase domestic access to electricity to 18% in 2017 and 32% in 2030.

The colossal Inga dam project consists of eight separate independent phases and upon completion has an energy potential of 100,000 MW. The Inga Dam project represents 40% of Africa's hydroelectric potential and 13% of the global potential. The construction of the Great Inga Dam on the Congo River basin, currently in phase three, produces 4,800 megawatts (MW). Once complete, it will generate enough electricity to power the country and others in the African continent. South Africa has been a major investor in the project and has plans to export half of the energy produced to meet its own demand. The DRC is hoping that the completion of the Great Inga Dam will provide a sustainable, reliable, and cost-effective energy source that can meet both domestic and intercontinental energy demands.

The DRC's energy potential is massive and government officials are aware of this. New regulations that focus on liberalization, or the loosening of restrictions in the energy sector, have allowed the DRC to secure investments that focus predominantly on energy production, more specifically hydroelectric energy. The government has identified 780 sites as having potential to generate hydropower. DRC's Prime Minister Matata advertises projects in the country as being extremely cost-effective and extraordinarily profitable. This and many other similar projects are going to shape energy innovations and investments not only in the DRC but in much of the African Continent for years to come.

The World Folio
http://www.theworldfolio.com/news/a-solution-to-africas-power-deficit/4242/

Apple's Successful Rebranding as a Green Energy Company

by Byron R. Núñez

In 2006, Apple Inc. ranked last in the Green Guide to Greener Electronics. The guide evaluates leading consumer electronic companies based on their commitment and progress in three environmental criteria: energy and climate, greener products, and sustainable operations. As Apple's global presence has grown, environmental groups and activists have been more critical of the company's energy suppliers. Today, the company runs entirely on renewable energy in 23 countries and is on track to achieve its long-term target of operating and manufacturing on 100% renewable energy.

On March 8, 2017, Apple moved one step closer by announcing that it is extending this goal to suppliers in Japan through its component supplier Ibiden. To meet its commitment, Ibiden will invest in more than 20 new renewable energy facilities, including one of the largest floating solar photovoltaic systems in the country. Ibiden's products help bring together the integrated circuitry and chip packages in Apple devices. Their renewable energy projects will produce over 12 megawatts of solar power, more than enough to power Ibiden's manufacturing while also providing support to Japan's nationwide efforts to limit carbon emissions.

Apple's decade-long commitment to using renewable energy for its own manufacturing and operations has had a global impact. In its 2016 supplier responsibility report, the company said that its energy efficiency program has reduced carbon emissions by more than 13,800 metric tons, and that its manufacturing partners will be generating over 2.5 billion kilowatt hours per year by the end of 2018. Lisa Jackson, the company's Vice President for Environment, Policy and Social initiatives, states, "as we continue our push to power our global operations with 100% renewable energy, it is more important than ever that we help our manufacturing partners make the same transition to cleaner sources, and set an example for other companies to follow." Apple's 10-year transformation from one of the worst green energy companies to one of the best provides a model for other companies and encourages them to do the same.

Apple Newsroom
http://www.apple.com/newsroom/2017/03/apple-takes-supplier-clean-energy-program-to-japan.html

Renewable Energy, a Success in Honduras?
by Byron R. Núñez

Climate change has had a tremendous impact on most Central American nations. The Centre for Research on the Epidemiology of Disasters showed that, in the last four decades, this area of the world has experienced a tenfold increase in extreme heat, drought, forest fires, storms, and floods. Researchers stated that countries like Honduras, Guatemala, and El Salvador are some of the most impacted since millions lack reliable food due to drought-related crop failures. As part of the Paris agreement that is set to go in effect in 2020, Honduras is investing in renewable sources of energy that can help prevent average global temperatures from rising above 2°C.

Earlier this year, advisor to the Honduran Council of Private Enterprise (COHEP), Solomon Ordonez, announced that the construction for the Platanares geothermal power plant in Honduras had reached an advanced stage. The $200 million geothermal project is funded by the Honduran government and foreign investment from the renewable energies company Ormat Technology Inc. The construction of the 35-megawatt project began in 2013 and is projected to generate an average annual revenue of $33 million.

Though renewable energy projects seem inherently progressive, they often displace indigenous groups from their land and interrupt their traditions and livelihoods. The area where the Platanares geothermal plant was constructed is home to the largest indigenous group of Honduras – the Lenca. The Centre for Research on the Epidemiology of Disasters reached out to 50 renewable energy companies and asked about their approach to human rights; only five said that they would commit to following the internationally recognized standards established in the United Nation's Declaration on the Rights of Indigenous Peoples, Ormat Technology Inc. was not one of them. Renewable energy in Honduras has been a success but it has come at the cost of displacing indigenous people to make space for these projects.

Quartz
https://qz.com/845206/renewable-energy-human-rights-violations/
Think GeoEnergy
http://www.thinkgeoenergy.com/35-mw-platanares-geothermal-plant-in-honduras-to-start-operation-in-2017/
United Nations Framework Convention on Climate Change
http://www.hcn.org/articles/you-cant-stop-illegal-immigration-while-denying-climate-change
http://newsroom.unfccc.int/unfccc-newsroom/honduras-submits-its-climate-action-plan-ahead-of-2015-paris-agreement/
TAGS: Geothermal Energy, indigenous people, Ormat Technologies Inc., Centre for Research on the Epidemiology of Disasters, the Paris Agreement, Lenca people.
https://www.facebook.com/thinkgeoenergy, https://twitter.com/thinkgeoenergy, https://www.linkedin.com/company/thinkgeoenergy

Brexit and Challenges to Unrenewable Energy—One Way National Policy Affects Renewable Energy

by Nadja Redmond

Recently, additions to the Brexit bill were denied by a vote in the British government. The conditions that were denied would've forced the government to consult Parliament on the permanency of contracts struck with the EU before the bill is finalized. Further issues having to do with the bill, such as treatment of EU-wide contracts, will undergo vote by Members of Parliament later this week. If this bill, which would trigger Article 50, is confirmed, the UK will commence with separation from the European Union.

One issue that remains up in the air before voting is finalized is treatment of previous agreements made within Euratom, the European Atomic Energy Community. Established in 1957, the international coalition intends to create and maintain a specialized market for nuclear power. There are currently plans to build a nuclear plant on Anglesey, an island near the coast of Wales. Dr. Glyn O. Phillips has raised concerns over the difficulties of procuring staff for the project if Brexit prompts a departure from Euratom. Phillips says a withdrawal would "be destructive to any nuclear work in the UK." This is mostly because European resources have been centered in Geneva, Switzerland.

While Professor Phillips and Horizon Nuclear Power, who is developing the Wylfa Newdd plant, are not necessarily optimistic about the process of hiring and training new employees, both have come to terms with the fact that leaving the European Union means also withdrawing from Euratom, as the two are legally tied. With Brexit

causing this type of separation in the nuclear energy industry, I wonder how renewable energy contracts are also being disrupted or supported?
http://www.bbc.com/news/uk-wales-north-west-wales-38884641
https://en.wikipedia.org/wiki/European_Atomic_Energy_Community
http://www.bbc.com/news/uk-politics-38895007.

Washington Clean Energy Testbeds
by Nadja Redmond

In Seattle, the newly developed Washington Clean Energy Testbeds have already gained a long waiting list of independent researchers and start-up managers who want to use the "clean energy innovation" facilities to ultimately revolutionize their research and business. The lab opened in late February, and these researchers will use the lab to creatively develop clean energy technology for use in their various projects. The facility aims to provide space and supplies for successful testing, scaling, and completion of these technologies. Funds for the 15,000-square foot facility came from the state by way of the University of Washington's Clean Energy Institute.

Inhabiting a space formerly used as a sheet metal manufacturer, the Clean Energy Testbeds will provide a common space for collaboration, use of expensive, powerful tools, and development of groundbreaking work. Users can rent time, lab space or technical expertise to test batteries, make cheap, paper-thin solar cell materials, or print high-precision screens, to name a few upcoming uses for the lab. Daniel Shwartz, a UW chemical engineering professor stresses how the lab will bring the many diverse projects happening on the UW campus into a personal lab, with opportunities to then scale the product or service to a larger market. "Too many start-ups have great ideas, but fail before fully demonstrating their technology. Amazingly, lack of easy access to facilities and expertise is often a barrier for big companies, too."

In relation to the current administration, lack of federal funding for this project is possible both because Seattle is a sanctuary city and because of Trump's plans to scale back the Climate Action Plan. Despite this, Governor Jay Inslee says he wouldn't let those changes interfere with the state's commitment to the clean energy industry. The implementation of the rent system in the lab is one method of combating worries about funding that is already in the works for the lab, and the CEI directors are in the process of devising more such methods.

[http://www.seattlepi.com/local/article/What-Trump-means-for-UW-s-new-clean-energy-10939078.phrp].
[http://www.pennenergy.com/articles/pennenergy/2017/02/ renewable-energy-seattle-lab-aims-to-speed-development-of-clean-energy-tech.html]
[http://www.pennenergy.com/articles/pennenergy/2017/02/renewable-energy-seattle-lab-aims-to-speed-development-of-clean-energy-tech.html]
[http://www.seattlepi.com/local/article/What-Trump-means-for-UW-s-new-clean-energy-10939078.php

Amazon Will Cover 50 Fulfillment Centers with Solar Panels by 2020

by Nadja Redmond

This year, Amazon plans to install solar panel generating systems on 15 fulfillment centers around the country, with plans to cover 50 similar national and international facilities by the year 2020. Already, the 1.1 million-square-foot rooftop of the fulfillment center in Patterson, California has these panels covering more than three-fourths of the space. These warehouses have a large footprint, so the implementation of these solar panels will put the large, unused space in the roofs of these warehouses to good use. The company has about 135 distribution and regional sortation facilities in the United States, and about 150 outside of the country. The solar panels provide 80% of power for covered warehouses, a significant dent in Amazon's overall energy consumption.

Many huge corporations with global presences are taking the step to embrace green power. This move for Amazon is especially important as well; the company has been criticized by Greenpeace for keeping information about its energy footprint under wraps amid its vast and quick growth. Other past initiatives by the company to increase use of renewable energy include a 2016 effort to power the data centers that house its cloud business by wind and solar energy.

Amazon's director of sustainability, Kara Hurst, talked about the reason this move to solar energy is possible: "Certainly the cost of the technology and the increase in availability has been a contributor." The facilities to receive the solar treatment this year are in California, Maryland, Nevada, Delaware, and New Jersey. These installs alone will generate up to 41 megawatts (MW) of electricity for these warehouses. For reference, 1 megawatt is estimated to have the capacity to power 164 houses. With these numbers, Amazon is set to become the number 7 corporate user of solar power.

http://www.seattletimes.com/business/amazon/amazoncom-plans-big-solar-power-rollout-at-warehouses/

What do Americans Think about Climate Change?

by Nadja Redmond

A recent New York Times article and Yale study revealed current thought trends among Americans regarding climate change. Overall, most Americans believe that global warming is happening, and that carbon emissions should be scaled back. Fewer are convinced these changes will affect them personally, however. A study released by the Yale Program on Climate Change Communication reveals more details.

On average, 69% of Americans (per congressional district) want to restrict carbon emissions from coal power plants, while the White House and Congress is prepared to do the opposite. Interestingly, the populations with more districts in which more than 80% of adults support strict CO_2 limits are focused on either coast. While most adults generally support limiting carbon dioxide emissions, members of Congress agree with the current administration and will most likely move to reverse Obama's plan to reduce greenhouse gas emissions.

Nationally, seven out of ten Americans support regulating carbon pollution specifically from power plants, and 75% support general CO2 regulation. Despite this, lawmakers are moving in the opposite direction. Somewhat counterproductive, though, is the belief most Americans have that climate change will harm Americans, but not them personally. While the former group is about 60% of Americans, the latter is only 30%. This is due to a lack of risk perception, and an issue that will have horrible long-term consequences. The NY Times article points out how humans are hardwired to react quickly and efficiently to immediate threats, but not motivated in the same way towards slow moving problems. This fact also affects the way Americans discuss climate change, with more in the West, which has experiences drought and wildfires, bring up the topic among family and friends.

For example, Texas overall is worried in different degrees over climate change, views differing based on demographics and age. The state has dealt with shifting weather patterns, rising temperatures, coastal hurricanes and western droughts for a while, which explains the higher than average degree of concern with climate change the southern part of the state has in relation to the surrounding areas.
[http://nyti.ms/2ntY5qT]

Honduras: Violence and Environmental Progress

by Yerika Reyes

In 2016, Empresa Nacional de Energía Eléctrica (ENEE), the Honduran state power company reported that 10.2% of energy in the country's electrical system was generated from the photovoltaic (PV) power plants. Honduras is the first non-island nation in the world to achieve a 10% share of solar energy in its national electricity assembly. Honduras also tops the charts as the country with the most installed PV capacity in Central America, with 433 MW of solar installed by the end of 2016, and is second in the whole of Latin America behind Chile, where more than 1 GW of PV has been installed. Public company ENEE, whose electric system owns nearly all the system in Honduras, reports that last year its plants added 8,673 gigawatt-hours (GWh) of generation that was purchased privately. Of that, 885 GWh corresponded to purchases from PV plants. Overall, renewable energy accounted for 52% of the Honduran power grid.

For all its environmental progress, high-ranking officials in the Honduran governments and business magnates are implicated in a surge of violence against environmental activists. An investigation by

Global Witness, an anti-corruption group, claims that the Honduran elites are using illegal means to terrorize communities with impunity. At least 123 land and environmental activists have been murdered in Honduras since a military coup d'état forced out the populist president Manuel Zelaya. The victims of these attacks have been members of indigenous and rural communities opposing mega-projects on their territories.

Last year Berta Cáceres, an indigenous activist was murdered. Her death triggered international denunciation but failed to stop the violence. Cáceres was actively campaigning to stop the internationally funded Agua Zarca hydroelectric dam on the Gualcarque river, which is sacred to the Lenca people. Prior to her murder, she had received years of death threats and state action because of her efforts. Since her death two of her colleagues have been killed. The investigation done by Global Witness revealed that the private company behind the dam, Desarrollos Energeticos SA (Desa), has a board of directors of influential political, military, and business leaders. For example, the company president, Roberto David Castillo Mejía, is a former military intelligence officer and employee of the Honduran state-owned energy company. The company secretary, Roberto Pacheco Reyes, is a former justice minister, while the company vice-president, Jacobo Nicolás Atala Zablah, is president of the BAC Honduras bank, and a member of a powerful business family. Additionally, an investigation by the Guardian revealed that Cáceres's name appeared with dozens of social activists on a military hit list assigned to US-trained special forces units.

This is not new for Honduras and environmental activists. The government has made environmentally destructive mining, agribusiness, tourism and energy projects the cornerstone of the country's economic growth strategy since the 2009 coup d'état. There has been an increase in violence against environmental activists who are trying to preserve their land. Environmental checks and balances have been ignored. The country's progress in renewables and solar are marked by violence against its own people.

https://www.pv-magazine.com/2017/01/30/honduras-first-country-in-the-world-with-10-of-solar-in-its-electricity-mix/
https://www.outsideonline.com/2136786/most-dangerous-place-earth-be-environmentalist
https://news.mongabay.com/2017/02/honduran-politicians-u-s-aid-implicated-in-killings-of-environmentalists/
"@USAmbHonduras Honduras is the first nation with 10% #solar, however it's in the aim for progress there has also been the mudering of indigenous activists for their land"

Model Region in China Aids in Solving National Challenge of Mismanaged Energy

by Mary-Catherine Riley

In July of 2016, China's National Administration announced that Jing-Jin-Ji would become a model region for energy reformation. Regional hindrances faced here mimic larger challenges China confronts. Jing-Jin-Ji is a highly polluted region even though it is located in an area that processes renewable energy sources. Challenges in this region also include that the government sets the prices of renewable energy, not the market, which causes artificially high costs for renewable energy. Furthermore, this pilot region faces troubles in energy integration where there is excess energy in some areas and not enough in others. Solutions can be found analyzing how locations with similar challenges have applied the best policies to solve regional issues. Solutions include this region adopting a spot power market which allows for daily energy transactions as opposed to the pre-allocated energy sources and quantities purchased months or even years prior.

Another large source of contention is that national policies have shifted towards the expansion of renewable energies, but the investment in renewable energy companies continues to grow. These competing policies and practices are a result of the mismanagement of energy. Renewable power plants are being constructed in areas where coal power is already meeting the needs of the area; however, that excess nonrenewable energy is not reaching the regions with unmet demand. Better coordination between plants and cities would ease this inconsistency. These solutions resolved in the Jing-Jin-Ji region would aid in solving national challenges. More specifically, the model begins to resolve the larger concern of how China wastes its manufactured renewable energy.

China serves as the leading investor in renewable energies but is also the leading nation in wasted, manufactured renewable energy. Unused Chinese wind energy averages at 21% nationally, soaring to 40% in some provences. The Paulson Institute, a nonpartisan think tank specializing in economic and environmental issues in China and the United States, calculates that China should focus on three elements to best reach its energy goals: China should allow the energy prices to be set by the market, focus on the Jing-Jin-Ji region in order to gain a better grasp of regional issues to apply the model nationally, and finally, China should establish a diligent timeline to achieve the ultimate goal of prioritizing renewable energy over coal. With such an extensive investment in renewable energy, this guideline should insure a valuable return.

(http://blogs.wsj.com/experts/2016/09/13/how-china-wastes-its-renewable-energy/).
(http://www.paulsoninstitute.org/paulson-blog/2016/07/29/a-step-forward-for-cleaner-energy-in-chinas-jing-jin-ji/).
Gallegos, Demetria, and Anders Hove. "How China Wastes Its Renewable Energy." The Wall Street Journal. Dow Jones & Company, 14 Sept. 2016. Web. 31 Jan. 2017.
Hove, Anders. "A Step Forward for Cleaner Energy in China's Jing-Jin-Ji." Paulson Institute. The Paulson Institute, 29 July 2016. Web. 31 Jan. 2017.

Former US President Carter Donates Land for Solar Panels to Give Energy to half his Town
by Mary-Catherine Riley

Former US President Jimmy Carter recently donated 10 acres of land in Plains, Georgia to SolAmerica to construct a 1.3 MW solar array.

Once constructed, the plant will provide 55 million kilowatt hours annually. An amount that will power 250 homes, encompassing over half the town's population. Georgia Power has signed on to a 25-year Power Purchase Agreement. This agreement solidified the movement, and construction of the solar plant began in October of 2016.

This effort is part of Former President Carter's lifelong commitment to solar energy. During his presidency in 1979, Carter installed 32 solar panels and a solar water heater at the White House in response to the Arab Oil Embargo. This action served to set an example to the American people about how to live sustainably and to bring awareness to the increasing capabilities of solar. The panels were removed during the Reagan Administration. Only recently, during the Obama Administration, were solar panels re-installed on the White House to produce 6.3 kilowatts of energy.

Land donation is generous but most times impractical. To combat this reality, there are local initiatives to provide income for those who lease out their lands to build solar farms. Companies such as Cypress Creek Renewables offer long-term passive income, lower operating costs, and reduced tax liability to those who lease their land to be used for solar panels. This method is yet another way to grow and incentivize the use of solar.

(https://cleantechnica.com/2017/02/14/jimmy-carter-builds-1-3-megawatt-solar-farm-hometown-plains-georgia/).

(http://www.ecowatch.com/jimmy-carter-solar-farm-2253275469.html, http://www.ajc.com/news/national/jimmy-carter-leases-his-land-solar-power-his-hometown/wYV7BLh5dgJQKUvV0czMZO/

(https://ccrenew.com/partners/landowners/).

Hanley, Steve. "Jimmy Carter Builds 1.3 Megawatt Solar Farm For His Hometown Of Plains, Georgia." CleanTechnica. Sustainable Enterprises Media, Inc., 14 Feb. 2017. Web. 24 Feb. 2017.

Jill Vejnoska Atlanta Journal-Constitution 11:40 A.m. Sunday, Feb. 12, 2017 National/World News. "Jimmy Carter Leases His Land to Solar Power His Hometown." AJC. Cox Media Group, 12 Feb. 2017. Web. 24 Feb. 2017.

"Landowners - Generate Passive Income With A Solar Farm." Cypress Creek. Cypress Creek Renewables, 2017. Web. 24 Feb. 2017.

Twitter Helping Solve South Australia's National Energy Crisis

by Mary-Catherine Riley

One million seven hundred thousand South Australians are experiencing a severe energy crisis after months of rolling black outs. In September, much of the state was left without power after a storm damaged crucial power lines. Another major blackout occurred in February after a heat wave incited an energy demand. Insufficient storage systems for renewable energies have led to this crisis. Without immediate action to the Hazelwood power station, New South Whales, Victoria, and South Australia have a 75% chance of blackouts next summer which would cost tens and tens of billions of dollars in the food, medicine, and processing industries. Furthermore, without government action, it will be impossible to have a constant gas supply through the winters of 2018 and 2019.

Solutions include solar or gas power plants and a pumped hydro project. SolarReserve proposes to build a solar thermal tower for $650 million while Reach Solar Energy bids a $660-million-dollar solar array with an energy storage system built into development for 100MW.

However, Elon Musk, co-founder of Tesla, offered to provide a solution to the crisis in a tweet to billionaire tech entrepreneur Mike Cannon-Brookes, stating "Tesla will get the system installed and working 100 days from contact signature or it is free." Later, Tesla solidified a price of $25 million for 100 megawatt hours of battery energy storage.

Pushback is coming from citizens frustrated that this crisis was only acted upon once two tech billionaires addressed the issue. Cannon-Brooks acknowledges this, but states that he is using his influence to ignite the conversation and initiate political support and funding.

Tesla's offer is tempting as it is cheaper and less risky than other bids; however, there is hesitation about a foreign company solving the problem when Australian companies could provide the same service while stimulating their own economy.

It will be interesting to see how it all plays out.

(http://www.theaustralian.com.au/business/opinion/robert-gottliebsen/energy-crisis-will-be-worse-than-expected-with-costly-blackouts-coming/news-story/8fd36298d245df103f8b0337056af04d).

(https://www.theguardian.com/sustainable-business/2017/mar/20/elon-musk-port-augusta-four-renewable-energy-projects).

http://money.cnn.com/2017/03/10/technology/elon-musk-australia-energy/).

Gottliebsen, Robert. "Energy Crisis Will Be Worse than Expected, with Costly Blackouts Coming." The Australian. The Australian, 20 Mar. 2017. Web. 22 Mar. 2017.

Kottasova, Ivana. "Musk Says He'll Fix South Australia's Energy Crisis in 100 Days -- or It's Free." CNNMoney. Cable News Network, 10 Mar. 2017. Web. 22 Mar. 2017.

Opray, Max. "Elon Musk, Meet Port Augusta: Four Renewable Energy Projects Ready to Go." The Guardian. Guardian News and Media, 19 Mar. 2017. Web. 22 Mar. 2017.

The Military is Pursuing Renewable Energy, But Why?

by Mary-Catherine Riley

The United States military is pursuing renewable energy, but why?

Colonel Brian Magnuson, the director of the expeditionary energy office in the Marine Crops states "There's a perception that the initiatives have to do with something other than extending our combat effectiveness." Former Assistant Secretary for Installations, Environment and Logistics of the Air Force under President George W. Bush, and other representatives from the Navy, Air Force and Army, echo his stance affirming "We're concerned about climate change... but the first mission is bombs on target... The Department of Defense is not the Department of Energy".

With this rhetoric in mind, however, the military branches are making considerable progress on their renewable energy sectors and here are the non-environmental reasons why.

There has been a considerable increase in power outages from 2000 to 2013. This is troubling for military bases particularly as they rely on antiquated infrastructure and the common electrical grid for power. In 2015, the Navy had 900 outages ranging from 15 hours on the East Coast and 30 hours in the Pacific. They are now implementing energy meters to regulate buildings' energy use over long periods of time to identify new efficient upgrades resulting in energy and cost savings and are investing in microgrids powered by renewables to generate energy for the bases. Moreover, the military could save over $1 billion and boost energy security by switching to microgrids and increasing energy efficiency.

Currently, the Army generates 12% of its energy renewably and has 17 large-scale renewable energy projects. In January of 2016, it signed the largest renewable energy project with Apex Clean Energy Inc. Their project is solar farm on Fort Hood that will generate 15 MW of power as well as an off-site wind farm that generates 50 MW of power.

So while climate change is not the first thing on the military's priority list, it is neat to see that there are other reasons to switch to nonrenewable energy.

(https://www.scientificamerican.com/article/energy-security-drives-u-s-military-to-renewables/).

(http://www.pewtrusts.org/en/research-and-analysis/analysis/2017/01/12/us-military-could-save-over-1-billion-and-boost-energy-security-new-research-finds).

(https://www.scientificamerican.com/article/energy-security-drives-u-s-military-to-renewables/).

Kaenel, Camille Von. "Energy Security Drives U.S. Military to Renewables." Scientific American. Scientific American, 16 Mar. 2016. Web. 08 Apr. 2017.

Swanson, Tom. "U.S. Military Could Save Over $1 Billion and Boost Energy Security, New Research Finds." The Pew Charitable Trusts. The Pew Charitable Trusts, 12 Jan. 2017. Web. 08 Apr. 2017.

The Caracol Industrial Park: Unfulfilled Energy Promises

by Sara R. Roschdi

The Clinton Foundation and The Inter-American Development Bank (IDB), have committed to funding over $300 million for the development of the Caracol Industrial Park. This Industrial park was built after the 2014 earthquake and was advertised to the world as a form of relief for the Haitian people. This project promised to support the development of the country's energy sector and the Inter-American Development Bank states they are committed to supporting the Caribbean and Latin America to "achieve universal energy access". Telesur reports that this over 600 acre industrial park has lead to the displacement of indigenous campesino land and the production of extremely low wage jobs. Lucas of CBC news reports, this industrial park promised to bring over 60,000 jobs to the region, yet has only provided roughly 5,500 minimum wage jobs with a daily salary of $5 US dollars and extremely poor working conditions (2014). The primary U.S. based employers in the Caracol Industrial Park are Walmart, Gap and Old Navy. There were over 3,500 campesinos evicted from their land, and these communities have lost their ability to sustain themselves and provide for the urgent need of sustainable food sources in the region. The IDB promised replacement land to the campesinos but none was provided. The United States Accountability Offices reports that USAID committed $170.3 million dollars to the creation of a power plant and $17 million of the budget has been used to create a plant with a capacity of 10 megawatts (2013). This power plant was anticipated to supply both the energy demands of the industrial park and expand to meet residential demands, but this promise is left unfulfilled. The United States, has donated two energy generators to supply the U.S. manufactures of the plant but these energy sources are not capable of meeting the energy needs of the Haitian people. The Haitian people are working to rebuild after decades of U.S. imperialism disguised as "development" and "aid" which is leading to the destabilizing of the Haitian government, and the exploitation of the country's people and natural resources.

Katz, Jonathan M. "A Glittering Industrial Park in Haiti Falls Short." Al Jazeera America. N.p., 10 Sept. 2013. Web. http://america.aljazeera.com/articles/2013/9/10/a-glittering-industrialparkfallsshortinhaiti.html

Kazi, By Nazia. "Haitian Campesinos Want Justice for Land Grab, Displacement." News | TeleSUR English. TeleSUR, n.d. Web. http://www.telesurtv.net/english/news/Haitian-Campesinos-Bring-Inter-American-Bank-to-Court-20170112-0036.html

Luksic, Nicola. "Haiti Shows How Wealthy Countries 'continue to Cause Disaster'." CBCnews. CBC/Radio Canada, 15 Sept. 2015. Web http://www.cbc.ca/news/world/haiti-shows-how-wealthy-countries-continue-to-cause-disaster-1.3228695

United States Government Accountability Office. HAITI RECONSTRUCTION USAID Infrastructure Projects Have Had Mixed Results and Face Sustainability Challenges (n.d.): n. pag., June 2013. Web. https://foreignaffairs.house.gov/files/zkVt_d13558._Restricted.pdf

The Bolivarian Revolution: "Sembrando Luz"
by Sara R. Roschdi

The Venezuelan Government States "Salvar el planeta es la meta superior" or "Saving the planet is the ultimate goal". As a part of Venezuela's initiative to end poverty, the country set out to provide energy to all. Foundation for the Development of Energy Services (Fundelec) provides a social program called Sembrando Luz or "Snow light" that brings solar power energy and water desalination systems to isolated regions of the country.

Fundelec's Sembrando Luz program has installed over 3,000 photovoltaic systems producing energy for more than 500,000 Venezuelans.

The program installs solar panels and teaches community members how to maintain them. As a requirement to receiving the solar panels, the community must be organized with a community council or a form of governance that holds responsibility for maintaining the solar electric facility, as a way to reduce the community's dependence on the federal government. Venezuela Analysis reports, that the Sembrando Luz program has two phases initiating with a team of engineers, electricians and national park workers that install a community solar panel to central community centers such as schools, council center, clinics and churches. In the second phase, each family home has a solar system installed with the goal of making the town 100% solar powered. This works to build self sustaining communities by providing the community with access to sustainable development. These solar power energy initiatives, allowed the community to spend more money on social programs such as meal programs, education and health services. It has allowed the community to better preserve foods and medical supplies, thus reducing waste, decreasing rates of malnutrition and infant mortality. The Sembrando Luz program is continuing the legacy of Hugo Chávez and keeping with the aspirations of the Bolivarian Revolution.

(http://www.fundelec.gob.ve/).

[http://www.lossinluzenlaprensa.com/a-traves-de-los-proyectos-sembrando-luz-y-mesas-tecnicas-de-energia-fundelec-ha-beneficiado-a-15-millones-de-venezolanos-con-sistemas-electricos/].

[https://venezuelanalysis.com/analysis/5598].

Santiago, Chile Takes a Stand on Climate Change

by Sara R. Roschdi

In June 2015, Santiago, Chile declared a state of emergency due to extremely poor air quality and high levels of toxic air pollutants. Pandey of the IBT reports that the emergency resulted in the temporary closing of over 900 industrial factories and the prohibiting of 40% of the 1.7 million cars to be on the roads. Santiago is located in a valley that makes it difficult for air to circulate. This traps harmful pollutants that settle within the city. These pollutants are reported to cause short-term respiratory problems, and long-term exposure are linked to increased rates of chronic bronchitis and poor lung function (Pandey, 2015). In May 2017, Chile plans to kick off their 10 year $US1 Billion plan to reduce air pollution Telesur reports that this plan strives to reduce greenhouse gas emissions by 60% The plan seeks to restrict the use of older less environmentally friendly vehicles during the day by only allowing them to operate during certain times of the day. This is comparable to programs in Mexico City and Quito, Ecuador that seek to reduce the number of cars on the road throughout the day to reduce traffic and environmental hazards. The Ministry of Environment reports that the plan also seeks to eliminate the burning of firewood by an estimated 20k households that use it for heating and cooking. This raises concerns for low income families that rely on the wood for heating and food preparation and the ministry responds that it will consider subsidizing alternatives to wood. The program will also increase regulations on commercial industries and ban agricultural burnings. Chile is leading the globe by taking an aggressive stance to reducing air pollution and committing to protecting the environment.

(http://www.ibtimes.com/santiago-smog-chile-declares-environmental-emergency-over-air-pollution-1976819).
http://www.telesurtv.net/english/news/Chile-to-Spend-US1-Billion-to-Battle-Pollution-in-Santiago-20161005-0007.html.

Paraguay's Export Energy Market

by Sara R. Roschdi

Paraguay is building an energy market that primarily functions as a low cost export for Brazilian industries. The Paraguayan President, Horacio Cartes, took office in 2013, and has rapidly been turning Paraguay into the "China" of Latin America. The business elite of Brazil and Argentina are outsourcing to Paraguay in order to gain access to low-wage labor and cheap energy. Paraguay produces much more energy than is needed and the primary source of energy comes from hydroelectric plants. Paraguay only consumes 10% of its energy with the other 90% going to the over 100 Brazilian industries within the country. Reuters Business news reports, that over 90% of the manufactured goods made in Paraguay go to Brazil. Cartes implemented a reform that reduced the cost to foreign exporters through tax reductions that charges exporters less than 10% and relieves them from the cost of custom tariffs. These neo-liberal policies are privatizing the country, diminishing funding for social programs, and creating a surplus of low-wage jobs. On February 13, Campesinos protested against these tax reliefs and demanded the resignation of President Horacio Cartes. Leader of the Paraguay Pyahura Party, Eduardo Ojeda, states, "Almost 1 million Paraguayans will not be having dinner tonight, while thousands are on the edge of death in the various hospitals of the country, because there are no drugs, no specialists, no infrastructure; thousands of campesinos are being evicted today from their lands by Brazilian soy producers supported by the national police and President Cartes" (Telesur, 2017). This increase in the exportation of the energy market, tax reliefs, and the growing surplus of low wage jobs is leading to the displacement of campesinos and indigenous communities. The international elite are profiting from the labor and energy of Paraguay, while the people are experiencing a diminishing quality of life.

http://www.telesurtv.net/english/news/Paraguays-Campesinos-March-to-Demand-Resignation-of-President-20170213-0023.html.

Ecuador Energy Sovereignty

by Sara R. Roschdi

The Ecuadorian Constitution of 2008 states the country's commitment to using eco friendly, clean and sustainable energy sources. They seek to attain energy sovereignty through producing sustainable energy sources that reduce their carbon footprint. This commitment seeks to restructure the energy dependency on the countries rich oil supplies. In August 2015, the Ecuadorian government dedicated US$7 billion to the production of hydropower, bioenergy, solar power, and wind power. Telesur reports, hydroelectric dams are expected to produce 2,827 megawatts of energy and reduce the CO2 emission by about 8.2 tons annually The 2012-2013 National Plan for Good Living sets a goal to reach 60% of the country energy sources to be renewable sources. The Ecuadorian constitution also includes the rights of nature and seeks to respect the rights of nature throughout it's development of renewable energy sources. They also seek to defend indigenous autonomy as they more forward with the building of hydroelectric dam. They are simultaneously implementing plans to protect the water and ecological reserves. This tasks developers and the government to build hydroelectric damns, protect the environment and defend indigenous autonomy, this has the potential to be an inspiration to fellow countries in the region. Water conservation is one of the top priorities of the National Strategy for Climate Change. Ecuador is investing 15 percent of its gross domestic product towards the production of clean energy sources. Energia16 reports, 8 dams were built from 2008-2016 with investments of US$11 billion leading up to production of 16% of the country's 2016 energy needs. The eight power plants are- Mazar Dudas, Toachi Pilaton, Minas-San Francisco, Coca Codo Sinclair, Manduriacu, Sopladora, Delsitanisagua and Quijos. Ecuador has more rivers than any other country and has the potential to have a flourishing hydroelectric energy market. The country is moving slowly as its priority is to respecting the rights of nature. Ecuador's energy market is primarily dominated by the oil rich resources of the country. This oil that can be traded on international markets and aid in the funding of social programs is being consumed by the country. Ecuador hopes to shift this and has an ambitious goal of having 94% of the country's energy needs met by environmentally sustainable and renewable energy sources by 2020.

(http://www.telesurtv.net/english/news/Ecuador-Moves-Toward-Hydroelectric-Energy-20150401-0029.html).

(http://www.energia16.com/hydroelectricity-the-key-for-the-ecuadorian-future/?lang=en).

Cuban Energy Revolution

by Sara R. Roschdi

In 2006 Cuba launched an "Energy Revolution" seeking to transform the country's energy market. In the initial stage of this revolution the country eliminated the country's reliance on incandescent light bulbs. Due to the effects of embargo and lack of international access to cleaner energy alternatives the country relied heavily on inefficient energy supplies such as crude oil, kerosene, and inefficient energy appliances. A government factory in Cuba produces 14,000 photovoltaic panels to support the growing solar power industry.

Cuba relies heavily on a "sweetheart" deal between Cuba and Venezuela where they receive a very low cost of oil. Unlike most countries currently Cuba is still dependent on crude oil for most of their energy needs. Cuba is known for old 1950's American cars that are notorious for their beauty, smog and large eco footprint. By 2030, Cuba plans to devote $3.5 billion to grow their renewable energy sources from 5% of the country's energy production to 24%. To meet this goal the country is developing their wind, solar, bioelectric and biofuel energy markets. The 20% plan seeks to grow the energy markets to 14% biomass, 6% wind, 3% solar and 1% hydropower. The country continues to grow this industry from international support from countries such as China and Russia who are planning to supply the wind turbines. Cuba's energy plan includes reducing the cost of delivered electricity from $21.10 kilowatt to $17.90 kilowatt by 2020. They plan to also reduce the country's grams of carbon dioxide emissions from 1,127 kilowatts to 1,018 kilowatts. Cuba is working to develop their renewable energy market and continue to develop industry within the country to grow this market. This is in a goal of reducing international dependency and fueling the country's economy.

(https://www.eenews.net/stories/1060020853).

(http://www.renewableenergyfocus.com/view/44708/cuba-seeks-to-expand-role-of-renewable-energy/)

(http://www.renewableenergyworld.com/articles/print/volume-12/issue-2/solar-energy/la-revolucion-energetica-cubas-energy-revolution.html).

http://www.renewableenergyworld.com/articles/print/volume-12/issue-2/solar-energy/la-revolucion-energetica-cubas-energy-revolution.html

http://www.eenews.net/stories/1060020853

http://www.huffingtonpost.com/david-sandalow/us-cuba-energy_b_9518226.html

http://www.renewableenergyfocus.com/view/44708/cuba-seeks-to-expand-role-of-renewable-energy/

Iran's Renewable Energy Future
by Sara R. Roschdi

Iran is known for its abundant oil supply, the US embargo and its alliance with leftist governments. Iran has a population of over 80 millions people that have been struggling with the contamination of their air due to climate change and the country's dependence on fossil fuels. Iran has the world's fourth-largest oil reserve and the second largest natural gas reserve. The U.S. embargo has reduced the country's access to refining and production equipment to produce oil that is safer to use as fuel. Due to climate change the traffic heavy capital city of Tehran has experienced a decrease in rainfall and strong winds which have led to an increase in smog and high air pollution levels. This air pollution crisis caused the country to go into a state of emergency in 2016 (Lila, 2016). This state of emergency led to the taking of over eight hundred thousand old heavy polluting vehicles off the road and a call for an increase use of public transportation as a way to reduce the number of vehicles on the road. Due to the effects of climate change the countries has developed a Five-Year Development Plan to develop renewable energy. This plan has been met with international attention for its development of nuclear reactors and energy. The Tehran Times reports that Iran has also been developing its wind, solar and geothermal energy capacities to currently resulting in 200 megawatts of environmentally sustainable energy. In 2014 with the start of this plan, President Rouhani's administration increased its investment in solar power projects from $12 billion dollars to $60 million dollars. With the lifting of sanctions against Iran they are seeking to receive $10 billion of international investments by 2018 and $60 billion by 2025 to grow their renewable energy markets. Iran is moving towards entering the global movement to save the environment and increase production of clean eco-friendly renewable energies.

(http://www.tehrantimes.com/news/300491/Iran-s-renewable-energy-market-blinks-at-investors).

(http://www.tehrantimes.com/news/300491/Iran-s-renewable-energy-market-blinks-at-investors).

(https://www.theguardian.com/world/iran-blog/2014/mar/10/irans-government-steps-up-efforts-to-tackle-pollution). (http://www.resilience.org/stories/2007-05-21/us-military-energy-consumption-facts-and-figures/%20%20http:/www.ucsusa.org/clean_vehicles/smart-transportation-solutions/us-military-oil-use.html#.WPRVzVPyt-U%20%20http://www.resilience.org/stories/2006-02-26).

http://www.cnn.com/2016/11/17/middleeast/tehran-smog-deaths-iran/

http://www.al-monitor.com/pulse/originals/2016/10/iran-renewable-energy-bushehr-wind-solar-development-plan.html

http://www.al-monitor.com/pulse/originals/2016/09/iran-tehran-water-crisis-water-pressure-drop-rationing.html

http://www.tehrantimes.com/news/300491/Iran-s-renewable-energy-market-blinks-at-investors

https://www.theguardian.com/world/iran-blog/2014/mar/10/irans-government-steps-up-efforts-to-tackle-pollution

http://www.thenational.ae/world/middle-east/iran-looks-to-solar-alternative-for-energy

Bolivia Renewable Energy Initiatives

by Sara R. Roschdi

Bolivia has introduced a new ministry of Energy to promote the government's "Programa Electricidad para Vivir con Dignidad" (PEVD) that promotes the country's transition to renewable energy. They are seeking to shift their dependency on oil to more eco friendly alternative energy sources. Castano of Renewable Energy World reports that Bolivia is seeking to increase its renewable energy market to account for 20%-25% of its national energy usage To achieve this ambitious goal Bolivia plans to develop the power projects that requires an investment of approximately US$30 billion. They plan to develop 34 electricity energy generation projects within the next ten years These energy projects include developments in biomass, solar, wind, geothermal and hydroelectric power. In the City of Santa Cruz, the wind energy potential is predicted to surpass countries such as Spain that are global leaders of wind energy production Hydrochina, a Chinese wind and hydro-power engineering firm, helped Bolivia open up its first wind farm in 2014 In 2015 the Ianas Energy Program reports that in the last 10 years rural area Bolivia have received 20,000 photovoltaic energy systems and 50 micro hydropower plants

(http://www.ianas.org/PDF/BolivaRenewableEnergyinBolivia1.pdf).
Bolivia is has some of the world's highest solar radiation and has the
potential to use this to become one of the leading solar energy
producers. In 2016 the Inter-American Development Bank invested
$100 million dollars for the "electrification" of rural areas. Currently
renewable energy accounts for 53% of all energy usage in these areas
Bolivia is developing its renewable energy sources in order to increase
access to energy for all the people of the country and as a step to
socializing the resources of the country. They are taking a step towards
increase energy access and respecting the rights of nature.

(http://www.renewableenergyworld.com/articles/2011/01/bolivia-plans-wind-power-other-
        renewable-energy-build-out.html).
(http://www.thinkgeoenergy.com/bolivia-creates-new-ministry-of-electricity-and-renewable-
        energy/).
(http://www.renewableenergyworld.com/articles/2011/01/bolivia-plans-wind-power-other-
        renewable-energy-build-out.html).
(http://www.iadb.org/en/news/news-releases/2016-09-02/rural-electricity-in-
        bolivia,11548.html).
(http://www.windpowermonthly.com/article/1174125/china-puts-bolivia-wind-map).

Energy Adaptations

Moving Forward with Plans to Build City of Floating Islands in Response to Climate Change

by Aurora Brachman

The California based nonprofit, the Seasteading Institute, announced plans to move forward with a project to build floating islands in the South Pacific. The group has raised $2.5 million thus far from 1,000 donors, and the government of French Polynesia has agreed to host the islands in a tropical lagoon. The group says the project could begin construction as early as next year pending the results of studies currently underway that assess the economic and environmental feasibility of the project. The concept of the floating islands is in reaction to the predicted displacement of people as a result of rising sea levels as a consequence of climate change. The most recent climate models predict the world's oceans may rise five to six feet by the end of this century. That is almost twice as much as the plausible worst case scenario put forth by the united nations in 2013. Despite plans to move forward, the project has many skeptics. The islands are projected to cost $10 to $50 million and house only a few dozen people. Given the price and how few individuals the islands could support, many doubt the projects feasibility on any kind of large scale and it does not appear to be a viable option for the Pacific Islanders most immediately affected by rising sea levels. And despite Polynesians and individuals from other poor and developing nations being the most vulnerable, the cost of living on these islands would only make them feasible for middle-income individuals from developed nations. This reflects the cruel irony that developed nations are the greatest contributors, the least affected, and the best equipped to respond to the effects of climate change. There are also concerns that this project perpetuates the tendency for technologically focused solutions for pacific islanders, and indigenous people more broadly, to ignore the needs of the Pacific Islanders themselves. Given the ultimate goal and magnitude of this project remains highly ambiguous, many have suggested this money could be better spent directed towards healthcare or education.

Ives, Mike. "As Climate Change Accelerates, Floating Cities Look Like Less of a Pipe Dream." Australia. The New York Times, 29 Jan. 2017. Web. 30 Jan. 2017. (https://www.nytimes.com/2017/01/27/world/australia/climate-change-floating-islands.html)

Riley, Kathleen. "Engineers are building a solar-powered, floating island city." Earth & Energy. Futurism, 19 Jan. 2017. Web. 30 Jan. 2017. (https://futurism.com/engineers-are-building-a-solar-powered-floating-island-city/)

Dubai Aims to be Global Leader in Climate Change Adaptation

by Aurora Brachman

At the 5th world government summit in Dubai the country outlined plans for how it will adapt to climate change, an endeavor which could give way to a multi-billion-dollar industry for the country. These plans have been put forward by the "Dubai Future Foundation", an organization established 18 months ago by the Prime Minister of the United Arab Emirates. Every major government agency is represented within the organization and the foundation's purpose is to anticipate future trends and find ways to capitalize and influence them.

The organization's leader says that while many countries are still debating whether climate change is even real, Dubai is fully acknowledging its reality and is trying to take proactive steps to combat and adapt to it. This commitment to combatting climate change is not only based in humanitarian and environmental concerns, but a desire to profit from being a leader in a budding industry. The organization anticipates that this industry could reap many billions of dollars for the country.

This commitment to combatting climate change may seem somewhat contradictory given that oil is one of the foundational components of Dubai's economy. However, Dubai has already begun to experience record-breaking summer heat and other countries in the region have experienced unrest as a result of food shortages caused by climate change. The government of Dubai recognizes that the era of oil is quickly coming to an end and that it has no choice but to find ways to adapt if it wants to continue to grow as a budding global power.

The election of Donald Trump, a vocal climate skeptic, has raised significant concerns for many in Dubai. Given the United States is the second largest polluter globally, their commitment to ending climate change is seen as essential. However, the UAE remains committed to combating climate change despite this disconcerting stance.

Pavlos, Zafiropoulos "Dubai's bid to cash in on climate change | DW Environment | DW.COM | 20.02.2017." DW.COM. N.p., 20 Feb. 2017. Web. 20 Feb. 2017. (http://www.dw.com/en/dubais-bid-to-cash-in-on-climate-change/a-37632398)

Kleiman, Joe. "Dubai's Museum of the Future Opens Exhibit About Climate Change Solutions at World Government Summit." InPark Magazine. InPark Magazine, 13 Feb. 2017. Web. 20 Feb. 2017. (http://www.inparkmagazine.com/dubai-museum-of-future-world-government-summit/)

New Evidence That Climate Change May be Causing the Global Jet Stream to Stagnate

by Aurora Brachman

One of the most troubling theories about the progression of climate change in the near future is the impact it may have on weather patterns. There is a belief that a warmer climate may cause planet-scale air patterns known as jet streams, which flow in waves across the globe, to stagnate. This means that weather patterns will last longer causing extreme droughts, heat waves, and excessive rainfall. A new study recently published in Nature supports this theory.

The study comes from Michael Mann of Pennsylvania State University as well as universities in Germany and the Netherlands. It found that in the summer the flow of the atmosphere is altering to the point where it is causing weather to get "stuck" more often. This freezing of weather patterns is likely because the Northern Hemisphere jet stream flows from west to east and is driven by the rotation of the earth as well as the differences between the temperature at the equator and the North Pole. The flow of the jet stream is stronger when the temperature difference is large and weaker when it is not. Because the arctic is beginning to warm as a result of global warming, the jet stream has become weaker and elongated.

There are, however, skeptics of this theory. Some fellow scientists who fully acknowledge the existence of climate change are hesitant to accept its findings. John Fyfe of the Canadian Center for Climate Modeling and Analysis at Environment and Climate Change Canada stated in response to this publication, "It is well established that there has been a human influence on the large-scale atmospheric circulation and temperature of the Northern Hemisphere. Mann et al. advance a theory that attempts to link those changes to changes in the spectrum of shorter-scale waves in the atmosphere, which are our weather makers. I do not believe that this theory is fully developed or that the implications have been fully explored, but I do think that Mann et al. study is a very good start."

Despite the skepticism over this particular finding, this study adds to the conversation that climate change is causing alterations in atmospheric flow, which is resulting in more extreme weather patterns. These Patterns are having effects that are already being experienced worldwide and it is important that we continue to take steps to better understand the effects of a warming globe.

Mooney, Chris. "One of the most troubling ideas about climate change just found new evidence in its favor." The Washington Post. WP Company, 27 Mar. 2017. Web. 28 Mar. 2017. (https://www.washingtonpost.com/news/energy-environment/wp/2017/03/27/one-of-the-most-troubling-ideas-about-climate-change-just-found-new-evidence-in-its-favor/?utm_term=.78cc1e9cf7a8)

Goenka, Himanshu. "Climate Change Impact On Jet Stream Leads To Extreme Weather Events, Study Finds." International Business Times. International Business Times , 28 Mar. 2017. Web. 28 Mar. 2017.

(http://www.ibtimes.com/climate-change-impact-jet-stream-leads-extreme-weather-events-study-finds-2516168)

Kickstarter Project Turns Air Pollution into High-Quality Ink

by Alejandra Chávez

Graviky Labs has found a way to filter black soot from car exhaust pipes and safely convert it into refillable, high-quality, water-resistant inks called Air-Ink. The project was curated by Anirudh Sharma, a computer engineering PhD student at the time, when he noticed the murky exhaust emitted by a passing bus while visiting his home in Mumbai. To achieve his vision, Sharma partnered with Nitesh Kadyan and Nikhil Kaushik. In the MIT Media Lab, they created a device, known as the Kaalink that captures soot particles before they are released into the air. The machine rids the soot of heavy metals and carcinogens and mixes the remaining purified, carbon-based pigment with vegetable oil. The amount of vegetable oil depends on the desired consistency of the pigment, which becomes pen ink, screen-printing ink, markers, oil-based paint, and outdoor paint. Cheaper carbon black inks are manufactured through the burning of fossil fuels, but Air-Ink hopes to combat that by scaling up production to "capture more pollution."

Some of the drawbacks of the project are that major companies would have to agree to use Kaalink devices in their vehicles, trucks, cranes, and other types of engines to operate at a full scale and meet demands. Partnership with major companies may lead to a spike in product costs. Regardless, as of February, 2017, the company has about 75 Kaalink devices in use and has launched a Kickstarter campaign to sell their products. One such product is their 30-milliliter pen, which contains about 45 minutes worth of air pollution. In total, the website claims that they have already "prevented the pollution of 1.6 trillion liters of air." Looking towards the future, the company acknowledges that Air-Ink "catches pollution on the back-end" instead of working to lessen fossil fuels, but they believe both efforts are needed.

Eldredge, Barbara. "Startup turns car exhaust into jet-black paint." Curbed. Vox Media, 9 Feb. 2017. Web. 28 Feb. 2017.
Curbed (http://www.curbed.com/2017/2/9/14555756/air-pollution-ink-pens-graviky-labs)
Graviky Labs. "AIR-INK: The world's first ink made out of air pollution." Kickstarter. Kickstarter PBC. Web. 28 Feb. 2017.
Kickstarter (https://www.kickstarter.com/projects/1295587226/air-ink-the-worlds-first-ink-made-out-of-air-pollu?token=05d1ee72)
Miller, Meg. "This MIT Spin-Off Turns Pollution Into Super-Black Paint." Co.Design. Fast Company, 7 Feb. 2017. Web. 28 Feb. 2017.
FostCo.Design (https://www.fastcodesign.com/3067701/this-mit-spin-off-turns-car-exhaust-into-super-black-paint)

As Climate Change Worsens, Mexico City Sinks

by Ethan Fukuto

Sinking at rates as much as 9 inches per year, Mexico City faces a looming threat of physically collapsing as the strains of climate change grow ever larger. Built upon clay lake beds and volcanic soil, Mexico City relies upon an increasingly difficult-to-extract supply of water beneath the city. The city's asphalt and concrete create a barrier between rain and the aquifers, lowering the water table and cracking the clay beds as the water drains, leading to uneven sinking throughout the city, made visible by tilting buildings, fissures, and cracks. As much as 40% of the city's water now comes from remote sources, though leaks in the pipes means that only around 60% of the water reaches its final destination.

Mexico City grew from about 30 square miles in 1950 to 3,000 in 60 years, causing increased strain on the city's quality of life. The city's Grand Canal, completed in the late 19th century to address the city's wastewater, was not built to address the specific set of problems now plaguing the city. Land subsidence has forced the government to install a series of pumps,which brings the water uphill and out of the city. Today it operates at about 30% capacity, and many residents are faced with unsafe, toxic water coming from their taps. Around 20% of residents cannot rely upon their tap water and thus many people hire trucks to bring in drinkable water. Residents are increasingly unable to survive: the New York Times references a study which predicts that 10% of Mexican citizens between ages 15 to 65 may emigrate north due to climate change and drought. Frank Biermann and Ingrid Boas, in advocating for action and protocol climate refugees, estimate that more than 200 million people worldwide may be forced to emigrate due to climate change by 2050. A series of interlocking issues, Mexico's water issues, compounded by climate change, requires immediate action and acknowledgement to ensure the safety of its 21 million residents.

Biermann, Frank, and Ingrid Boas. "Protecting climate refugees: the case for a global protocol." Environment: science and policy for sustainable development 50, no. 6 (2008): 8-17.

Kimmelman, Michael. "Mexico City, Parched and Sinking, Faces a Water Crisis". New York Times. 17 February 2017.
https://www.nytimes.com/interactive/2017/02/17/world/americas/mexico-city-sinking.html?rref=collection%2Fsectioncollection%2Fearth&action=click&content Collection=earth&region=rank&module=package&version=highlights&contentPlac ement=1&pgtype=sectionfront.

Zarembo, Alan. "The Canal From Hell". Newsweek. 20 May 2001.
http://www.newsweek.com/canal-hell-152561.

Wax Worm Caterpillar May Provide Key to Reducing Plastic Waste

by Ethan Fukuto

Researchers at the University of Cambridge have discovered that the larvae of wax moths, *Galleria mellonella*, can digest plastic. Federica Bertocchini, one of the lead researchers, and colleagues discovered that plastic bags holding wax worms were riddled with holes. Subsequent studies found that the worms can eat plastic, with holes forming in plastic samples in under an hour. Ninety-two percent of plastic production creates polyethylene and polypropylene. Polyethylene makes up about 40% of plastic demand, and is most commonly used in packaging. Only about 9.25% of polyethylene in the United States is recycled, with 75.5% ending up in landfills and 15% burned for energy.

Polyethylene also has very stable bonds, and does not easily biodegrade in landfills. But placing the worms on a commercial plastic bag for twelve hours, the researchers found that the worms were able to eat ninety-two milligrams of polyethylene. To determine whether or not the worms were simply eating the plastic rather than digesting it, they then blended together worms and placed the mixture of cells onto a polyethylene film. After fourteen hours, 13% of the polyethylene mass had been lost. Importantly, the researchers found that the mixture of cells from the wax worms are able to chemically transform the polyethylene in the plastic into ethylene glycol.

Since wax moths lay their eggs in beehives, their larvae naturally eat compounds with a similar chemical makeup of plastic, namely compounds with carbon-carbon bonds. Researchers are still unsure of the exact mechanisms at work in this process, though they believe microbes in the worm's gut may be breaking down the chemicals. Or insect itself may produce enzymes that contribute to breaking down plastic; further research is planned to determine the exact molecular processes. Discovering the details may provide the means to create a biotechnological process to cleaning up polyethylene in the environment.

Briggs, Helen. "Plastic-eating caterpillar could munch waste, scientists say." BBC. 24 April 2017. < http://www.bbc.com/news/science-environment-39694553>

Cell Press. "Wax worm caterpillar will eat plastic shopping bags: New solution to plastic waste?" ScienceDaily. ScienceDaily, 24 April 2017. <www.sciencedaily.com/releases/2017/04/170424141338.htm>.

Khan, Amina. "Stubborn plastic may have finally met its match: the hungry wax worm." LA Times. 24 April 2017. < http://www.latimes.com/science/sciencenow/la-sci-sn-worm-eats-plastic-20170424-story.html>

Eggshells and Tomatoes: Making Tires From Sustainable Alternatives

by Sagarika Gami

Researchers at Ohio State University have found that food waste can partially replace the petroleum-based filler that is most frequently used in manufacturing tires. This filler, carbon black, has been used for more than a century. American tire companies often purchase it from overseas. Approximately 30% of a typical car tire is made from carbon black, which is why tires appear black. Using this material makes the rubber durable, but carbon black is getting more and more difficult to come by. Katrina Cornish, Ohio Research Scholar and Endowed Chair in Biomaterials at Ohio State, has been working to cultivate new domestic rubber sources for years. She has recently found a method to turn eggshells and tomato peels into replacements for carbon black.

Cornish and her team are getting eggshells and other food waste from Ohio food producers. According to USDA, Americans consume nearly 100 billion eggs per year and 13 million tons of mostly canned or processed tomatoes per year. The eggshells cracked in commercial food factories are then sent to the landfills by the ton, but in the landfills, the mineral-packed shells can't break down. Similarly, tomato skins are most often discarded because commercial tomatoes have been bred to grow thick, fibrous skins to withstand being transported. Food companies that seek to make products out of tomatoes, like tomato sauce, peel and get rid of the skin because it is no longer easily digestible. Cindy Barrera, a postdoctoral researcher in Cornish's lab, explains, "Fillers generally make rubber stronger, but they also make it less flexible. We found that replacing different portions of carbon black with ground eggshells and tomato peels caused synergistic effects - for instance, enabling strong rubber to retain flexibility." This new rubber looks reddish brown, depending on how much eggshell and tomato are in it.

In the tests done so far, the rubber made with food waste fillers exceeds industrial standards for performance. This technology will work to solve three problems. First, rubber manufacturing will become more sustainable, second, American dependence on foreign oil will be reduced, and third, some proportion of waste will be kept out of the landfills. Ohio State has licensed the patent-pending technology to Cornish's company, EnergyEne.

Spacedaily.com
(http://www.spacedaily.com/reports/Turning_food_waste_into_tires_999.html)

Neuralink: Elon Musk, Helping us Keep up with Artificial Intelligence

by Sagarika Gami

Elon Musk is out with yet another business venture, this time connecting our brains to computers. Musk has founded Neuralink, a brain-computer interface venture, at its earliest stage of existence. He is working on creating devices that can be implanted in the human brain to help humans merge with software and keep up with advancements in Artificial Intelligence. He believes that humans merging with machines is the only way to stop us from becoming irrelevant in the world of AI.

These implants could work to improve memory and allow for more direct interfacing with computing devices; the goal is to facilitate a closer merger of biological intelligence and digital intelligence. The implants are a facet of "neural lace," short hand term for brain-computer interface, which will allow people to directly communicate with a computer without the need for a physical interface. Essentially, it is a series of electrodes implanted in our brain, which will then allow a wireless link to computers. Our thoughts could be uploaded and downloaded as desired. Musk explains the implant arrangement as a third digital layer, the limbic system and the cortex being the first two layers, all three layers working symbiotically together.

Currently, electrode arrays and implants have been used with patients of Parkinson's disease, epilepsy, and other neurodegenerative diseases. Very few people have implants inside their skulls because it's dangerous and invasive to operate on the brain in this manner.

Musk has said, "merging in some symbiotic way with digital intelligence revolves around eliminating the input/output constraint. So, it would be some sort of direct cortical interface." He adds that this "merging" will not require surgery, but instead could go through veins and arteries, as they provide a complete roadway to all of our neurons.

Theverge.com (http://www.theverge.com/2017/3/27/15077864/elon-musk-neuralink-brain-computer-interface-ai-cyborgs)

Npr.org (http://www.npr.org/sections/thetwo-way/2017/03/28/521763351/elon-musk-seen-targeting-human-computer-link-in-new-venture)

Pcmag.com (http://www.pcmag.com/news/352685/elon-musks-neuralink-will-connect-our-brains-to-computers)

Conceptualizing a Thermal Computer
by Sagarika Gami

Two engineers from the University of Nebraska-Lincoln are working to combat one of the biggest problems with computers–overheating. Instead of working to cool down computers from the heat, they are embracing it as an alternative energy source. The goal is to make for computing at very high temperatures. The research is developing a nano-thermal-mechanical device, known as a thermal diode. One of the engineers is Side Ndao, an assistant professor of mechanical and materials engineering and the other is Mahmoud Elzouka, a graduate student in mechanical and materials engineering.

Ndao's thought is that whatever is done with electricity should also be able to be done with heat because of their similarities. Both are energy carries, so if you could control heat, it could be used to do computing. So far, their devices have worked in temperatures that reach 630°F. This 'thermal computer' has many implications–it could be used in space exploration, oil drilling, exploring the earth's core, and more. Additionally, the thermal diode could help to limit how much energy is wasted, by using an energy source that has thus far been overlooked.

A study notes that about 60% of energy produced for consumption in the US is wasted in heat. If this heat can be utilized, both cost and waste will be brought down. There is a lot of further work to be done in making the device more efficient in order to do specific computations, which run an experimental logic system for proof-of-concept. The researchers have recently filed for a patent to improve the performance of the thermal diode. Ndao hopes to create the world's first working thermal computer and to explore the vast possibilities ahead.

Spacedaily.com

> (http://www.spacedaily.com/reports/Harnessing_heat_to_power_computers_999.html)

This App Takes Climate Change from a Theory to Reality

by Genna Gores

The Artist Justin Brice Guariglia celebrated Earth Day in 2017 by releasing his new app "After Ice," which allows its user to see if their current location will be underwater in the coming years due to climate change. in 2015, Guariglia started participating in survey flights with NASA scientists during Operation IceBridge, which looked at regions with melting polar ice caps, to better understand how these areas are affected by climate change. As an artist who typically works with concepts of science and philosophy, he wanted to take the big and theoretical aspect of climate change and make it something that any human could comprehend. Guariglia exemplifies this concept when he says, "'We're so disconnected from this [phenomenon], but when you experience it in augmented reality through an app, or even just walk down to the waterfront and recognize that that water comes from ice melting in Greenland, it becomes real.'" He hopes that with this app he can make climate change more accessible to people, and make it clear that sea levels rising is a reality that will happen for current generations.

When opening the application, a user is asked to give their location, and then they can see whether their current position will be underwater anytime between now and when the polar ice caps are fully melted. Eillie Anzilotti, a reporter from Fast Company using the app, found out that with 79% of total melt, her neighborhood in New York will be 208 feet underwater. She could see a picture of herself standing under the waves while fish swam by. After this section, the app asks you to take a selfie with yourself underwater and share it with your friends. Guariglia hopes that this kind of social media sharing will personalize the problem of sea levels rising, and will spread the word that this a pertinent and time-sensitive issue. The app then hits home with its final page showing the famous Bull of Wall Street under water in 2080, which is when the sea will rise by approximately six feet. While playing with this application does not solve the issue of climate change, hopefully it will make the concept real enough for the average person to take action.

[https://www.fastcompany.com/40411011/this-app-shows-you-what-life-will-be-like-when-the-worlds-ice-melts]

https://augmented.reality.news/news/after-ice-shows-what-climate-change-is-about-do-our-world-0177221/

https://www.fastcompany.com/40411011/this-app-shows-you-what-life-will-be-like-when-the-worlds-ice-melts

The Many Effects of Seagrass Disappearance
by Parker Head

Carl Zimmer examines the ecological and economic devastation incurred by the disappearance of seagrass. By examining this often-overlooked organism's role within the aquatic ecosystem, this article reveals how the human-based harm done to seagrass pastures is effecting fish and other wildlife, and if it continues, how it will begin to harm humanity.

Seagrass acts as a filtration system for the ocean. Every continent, except Antarctica, is ringed with vast prairies of seagrass. Seagrass releases so much oxygen that it can cause the surrounding water to bubble "like champagne". The oxygen produced by seagrass kills pathogens in the water that are harmful to humans and coral alike. Researchers studying the grass have discovered that a portion of oxygen produced is diverted to the roots, detoxifying the soil in which they grow; therefore, giving seagrass the ability to filter fertilizer runoff and other manmade pollutants out of the underwater ecosystem.

As filterers, they exist at the bottom of the ecological ladder, exerting a great effect on the countless organisms above them. A 2014 study by Cambridge University studied seagrass meadows in southern Australia in order to estimate their economic value for the fishing industry. It found that each acre of seagrass adds about $87,000 annually to the industry.

With the sprawling benefits of seagrass outlined, Zimmer goes on to report how their decline, and the subsequent cessation of these beneficial features, could soon begin affecting us. It is estimated that nearly 29% of the Chesapeake Bay's seagrass has been eradicated since 1991 due to global warming and increased pollutant runoff. With warmer water temperatures, the seagrass cannot produce as much oxygen, thereby allowing more toxins to remain in the water. The clouding of water due to fewer amounts of pollutants being stored and filtered through the soil by seagrass has raised average water temperatures, thereby compounding seagrass's inability to produce oxygen. This will not only effect humans financially – the Chesapeake Bay being a major fishing area, especially for blue crabs who use seagrass meadows as nurseries – but also physically. Zimmer writes, "As seagrass meadows disappear, that carbon is being released back into the ocean. Some of it may make its way into the atmosphere as heat-trapping carbon dioxide".

[https://www.nytimes.com/2017/02/16/science/seagrass-coral-reefs-pathogens-global-warming.html]

[https://www.nytimes.com/2017/02/16/science/seagrass-coral-reefs-pathogens-global-warming.html].

Zimmer, Carl. 2017. Disappearing Seagrass Protects Against Pathogens, Even Climate Change, Scientists Find. New York Times. February 16, 2017.
https://www.nytimes.com/2017/02/16/science/seagrass-coral-reefs-pathogens-global-warming.html?rref=collection%2Fsectioncollection%2Fearth&action=click&contentCollection=earth&region=rank&module=package&version=highlights&contentPlacement=1&pgtype=sectionfront

Lefcheck, Jonathan S. et al. Multiple stressors threaten the imperiled coastal foundation species eelgrass (Zostera marina) in Chesapeake Bay, USA. Global Change Biology. February 6, 2017.
http://onlinelibrary.wiley.com/doi/10.1111/gcb.13623/abstract

Blandon, Abigayil, and Philine S.E. zu Ermagassen.Quantitative estimate of commercial fish enhancement by seagrass habitat in southern Australia. Science Direct. December 5, 2014.
http://www.sciencedirect.com/science/article/pii/S0272771414000213

Effect of Runoff from Road Salt Alternatives on Aquatic Life

by Parker Head

Mary L. Martialay reports on the first study ever to examine the effects of road salt alternatives on aquatic ecosystems. The recent study, conducted by Rensselaer Polytechnic Institute, IBM Research, and the FUND for Lake George, and published in the *Journal of Applied Ecology*, analyzed the effects of varying concentrations of five different road salt treatments on artificially constructed freshwater ecosystems. The constructed aquatic ecosystems contained many organisms that are foundational to freshwater food web life, including algae, zooplankton, amphipods, isopods, and snails The five types of road salt studied were: sodium chloride (rock salt), magnesium chloride, magnesium chloride with a small amount of sodium chloride (e.g., Clear Lane®), sodium chloride mixed with beet juice (e.g., GeoMelt®), and magnesium chloride that was mixed with a distillation byproduct (e.g., Magic Salt®). The Clear Lane®, GeoMelt®, and Magic Salt® treatments are all marketed as environmentally friendly alternatives to standard rock salt because the additives allow for a fewer number of treatments to keep roads ice free. This study wanted to examine what effects the additives have on aquatic ecosystems, which are affected by the runoff.

The results revealed that the additives did indeed drastically alter the aquatic ecosystems they drained into, but not necessarily negatively. The alternative salt treatments increased algal growth and more than tripled the amphipod population. Overall, the organic additives were "like adding food to the lake," says Dr. Schuler, the first author on the paper. But, while this research shows that road salt alternatives do indeed alter aquatic ecosystems, it does not conclude whether this change should be regarded as good or bad. Dr. Reylea, another author of the paper, says, "Our research shows that these chemicals can cause changes to the food web, but we can't tell you whether that is desirable or not...More algae means more zooplankton and more fish, and the angler might like that. But more algae also means turbid water, and a homeowner may not like that"

[https://news.rpi.edu].

Martialay, Mary L. 2017 Road Salt Alternatives Alter Aquatic Ecosystems. RPI News. February 28, 2017.

Schuler, Matthew S. et al. How common road salts and organic additives alter freshwater food webs: in search of safer alternatives. Journal of Applied Ecology. February 28, 2017. http://onlinelibrary.wiley.com/doi/10.1111/1365-2664.12877/abstract

Journal of Applied Ecology Twitter: https://twitter.com/JAppliedEcology?ref_src=twsrc%5Egoogle%7Ctwcamp%5Eserp%7Ctwgr%5Eauthor

RPI News Twitter: https://twitter.com/RPInews

Sticky Gel Allows for the First Robotic Pollinator

by Parker Head

A recent paper published in the journal *Chem* reveals the results of the research team of Chechetka *et al.* efforts in fabricating a robotic pollinator.

The need for robotic pollinators has rapidly grown. The commercialization of honeybees has seen a dramatic increase in the last three decades. In 1998 2.5 million honeybee colonies were rented for pollination in the United States alone, an 18% increase from 1988, and this number has only continued to rise with the endangerment of the species. And the increased commercialization of honeybees has only quickened the pace of the decline in the honeybee population. A robotic pollinator is timely, as it would be economically profitable while also helping to relieve some of the burden of commercialization from the actual honeybees, promoting species regeneration.

But, a surrogate for the bees is not so simple to create. Honeybees are completely covered in microscopic hairs of varying shape. Different hair shapes collect and deposit specific pollens. The crucial component of the team's artificial pollinator is a sticky gel synthesized by Dr. Miyako, a chemist at the National Institute of Advanced Industrial Science and Technology in Japan. The research team applied the gel to horse hair and found that the combination proved very efficient at collecting pollen, acting similar to honeybee hair in its ability to pick up and deposit pollen. When a swatch of the gelled horse hair was attached to a drone the team could brush against plants and pollinate them as a honeybee would. Moving forward, the main obstacles to overcome are mechanical, as current micro-aviation vehicles require relatively large batteries. Ideally, future artificial pollinators will be more autonomous; instead of being controlled by an operator, they will be outfitted with sensors allowing them to detect plants via olfaction or color.

[http://www.cell.com/chem/fulltext/S2451-9294(17)30036-0]

Chechetka, Svetlana A. et al. 2017. Materially Engineered Artificial Pollinators. Chem. February 9, 2017.
    http://www.sciencedirect.com/science/article/pii/S2451929417300323
Mandelbaum, Ryan F. 2017. Scientists Are Building Bee-Like Drones to Fight the Coming Bee-Pocalypse. February 9, 2017.http://gizmodo.com/scientists-are-building-pollinator-drones-in-preparatio-1792171458
Cell Press' Twitter:
    https://twitter.com/CellCellPress?ref_src=twsrc%5Egoogle%7Ctwcamp%5Eserp%7Ctwgr%5Eauthor

Cool New Way to Recycle E-Waste

by Parker Head

It has become the norm for many people to upgrade their electronic devices once, or even twice, a year. And, with most people doing this for multiple devices, the amount of electronic waste deposited into landfills has increased dramatically. This trend has had an exceptionally detrimental impact on the environment and has given rise to the need for an innovative recycling method. Michael Irving reports on a new method being devised by researchers at Rice University and the Indian Institute of Science that involves freezing the e-waste with liquid nitrogen allowing for an almost waste-less recycle processing.

The environmental impact of e-waste is not only unprecedented in its volume, but also in its toxicity. Some common elements used in electronics include mercury, cadmium, and chromium. When electronics are thrown away in landfills, dangerous heavy metals like these leach into the soil, poisoning the ecosystem. There are also many valuable metals used in electronics such as gold, silver, and copper that would be economically advantageous to recycle instead of throwing out. But, e-waste proves especially difficult to recycle. Chandra Sekhar Tiwary, the lead researcher of the joint effort, says of the current inadequacies of recycling methods' treatment of e-waste, "burning or using chemicals takes a lot of energy while still leaving waste". These methods also tend to fuse the component metals, toxic and non-toxic alike.

Her team of researchers has therefore developed what they call a cryo-mill. The cryo-mill is composed of a canister containing a steel ball which is continually sprayed with liquid nitrogen while vibrating at very high speeds. The e-waste is placed inside the canister and pulverized into a nanodust. But, because of the extremely low temperature maintained during the grinding process, the nanodust is composed of discreet particles of the different metals and polymers. The different particle types separate when placed in a water bath, allowing for almost complete recycling.

http://newatlas.com/freezing-crushing-ewaste/48524/
Irving, Michael. 2017. E-Waste not, want not: Freezing, crushing and reclaiming old electronics. New Atlas. March 21, 2017. http://newatlas.com/freezing-crushing-ewaste/48524/
Kuehr, Ruediger et al. 2015. Discarded Kitchen, Laundry, Bathroom Equipment Comprises Over Half of World E-waste. United Nations. April 19, 2015.

https://unu.edu/media-relations/releases/discarded-kitchen-laundry-
bathroom-equipment-comprises-over-half-of-world-e-waste-unu-report.html#info

Amazon's Drone Delivery

by Cybele Kappos

Amazon is constantly striving to innovate its drone program. The latest is a technology that would release the package and deploy a parachute on the descent. Alex Hern explores the latest patents of Amazon's program in his 2017 article.

The delivery drones are not new, but the technology used to ensure the package's arrival is. Initially, the use of drones could only be done for customers who had large gardens. In seeking to address that problem, the drone would release the package from above and deploy a parachute to slow the descent in order to ensure the safe landing of the package contents. However, the parachute could cause problems if, for example, a strong gust of wind were to push it off course. The second aspect of this patent contains another innovation that addresses this potential problem. The company proposes that the drone continue to hover nearby and monitor the package's descent. If it moves off course, the drone can activate a variety of methods in order to rectify the descent. Some of these methods are flaps and bursts of compressed air.

Another aspect of the patent concentrates on how to ensure that the package is dropped without continuing on the same trajectory as the drone. The solution is to propel the package from the drone by applying force on to the package. This force can be applied in a number of ways, including pneumatic actuators, electromagnets, and spring coils. This is all in order to ensure a vertical descent path of the package. Amazon's latest patent for its drone program is an innovation that seeks to increase the number of customers suitable for its trials. There were only two members of the trial that involved the drone descending and landing with the package because a large garden was needed for the landing. This exciting patent demonstrates the company's willingness to refine the drone delivery system.

Hern, Alex. Amazon planning to use drones to drop parcels by parachute. 15 February, 2017
https://www.theguardian.com/technology/2017/feb/15/amazon-files-patent-parachute-aided-drone-delivery
http://www.usatoday.com/story/tech/news/2017/02/15/how-amazon-might-deliver-packages-drone/97937664/

Printable Electronics

by Cybele Kappos

Researchers at MIT have been experimenting with printable electronics. According to professor A. John Hart, there is a huge need for such things. Their utility is their low cost and their capacity for simple computation and interactive functions. Lee's 2016 article discusses how an MIT team has been exploring the advantages of nanotechnology in printing electronics.

Already some people have experimented with inkjet printing and rubber stamping, but the results have been inaccurate and difficult to work with on small scales. The complications involved fuzziness, coffee-ring patterns, and incomplete circuits. For electronics like a transistor or thin film, accuracy on a small scale is a very important characteristic. In response to these complications, MIT turned to nanotechnology. Their idea was to create a nanoporous stamp that would allow for a uniform distribution of nanoparticles, comprising polymer-coated aligned carbon nanotubes. They created the stamps by growing carbon nanotubes on different patterns on silicon. In addition, the researchers constructed a model to measure the amount of force in stamping, ensuring an equal amount of pressure. Pressure is an important criterion in ensuring a better result. Using all this information, they built a printing machine to mimic the industrial process and test their idea. The result was a much higher resolution stamp than other methods. They also found that the speed with which they print, at 200 millimeters per second, is competitive to industrial printing technologies. This speed in addition to the improved resolution is a great advancement in printable electronics.

The printed patterns are highly conductive and suitable for several applications, such as high-performance transparent electrodes. The stamp is able to print the electronic ink onto both flexible and rigid surfaces. According to Hart, the printing process "is an enabling technology for high-performance fully printed electronics."

The team will be looking to integrate their technology into 2D materials such as graphene.

[http://newatlas.com/ultrathin-high-res-electronic-decals-mit/46852/] Lee, Lisa-Ann. Turning the potential of electronic printables into a real breakthrough. December 9th, 2017
http://newatlas.com/ultrathin-high-res-electronic-decals-mit/46852/
http://advances.sciencemag.org/content/2/12/e1601660

Hacking with Sound Waves

by Cybele Kappos

Researchers discovered a security loophole that is more harmless than it seems. A group from the University of Michigan and the University of South Carolina demonstrated a vulnerability of accelerometers that would allow them to take control of the devices. Accelerometers are components of smartphones, fitness monitors, and even automobiles. Their experiment showed how they added fake steps to a Fitbit fitness monitor and played a music file from the speaker of the smartphone to control its accelerometer. This allowed them to interfere with the software of the phone, such as an app that controls a toy car. The researchers described the process as parallel to an opera singer who hits a note to break a wine glass. Similarly, this vulnerability allows them to enter commands. They discovered the flaw in more than half of the 20 commercial accelerometers from five brands

229

they tested. It is another challenge of the security of technology that is constantly spreading around the world. In relation to the rapidly advancing future of self-driving cars and trucks, the possibility of attackers remotely controlling vehicles is disconcerting. The computer security researchers did establish that the discovery was not critical. It reveals the "cybersecurity challenges inherent in complex systems in which analog and digital components can interact in unexpected ways."

Accelerometers measure acceleration and are manufactured as silicon chip-based devices. They are used for navigating, for detecting the direction of a tablet computer and for fitness monitors. The example of the toy car was insignificant but it points to other possibilities such as controlling the insulin dosage of a diabetic patient or tampering with the heart rhythms of a pacemaker.

In the researchers' experimentation, 75 percent of the devices were affected by the test and 65 percent were controlled by the test. The Department of Homeland Security is expected to issue a security advisory for the manufacturing of these chips by these companies.

Markoff, John. It's Possible to Hack a Phone With Sound Waves, Researchers Show. March 14, 2017
https://www.nytimes.com/2017/03/14/technology/phone-hacking-sound-waves.html
https://www.nytimes.com/2017/03/14/technology/phone-hacking-sound-waves.html?rref=collection%2Fsectioncollection%2Fscience&action=click&contentCollection=science&region=stream&module=stream_unit&version=latest&contentPlacement=5&pgtype=sectionfront
http://gizmodo.com/hackers-can-now-use-sound-waves-to-take-control-of-your-1793259066

Ethical Chicken Nuggets

by Cybele Kappos

Chicken and duck made in a lab from poultry cells has just been served at Memphis Meats. This is a very important development towards solving the problems of the modern meat industry. Irving's article talks about the success of this meal not only in terms of ethics but in terms of taste. The fact that the meat was grown from poultry cells means that no animals were harmed in the process of making the meals. In addition to the treatment of the animals, the meat industry has a severe environmental impact. Feeding, breeding and keeping livestock contribute to this problem as the animals require vast areas of land and large amounts of food, water and care. In addition, they release greenhouse gases themselves. In 2013, people tasted beef that was lab-grown but according to reports, the meat was bland and extremely expensive to make. Other attempts were made by the company Impossible Burger, but Memphis Meats just unveiled their southern fried chicken and duck à l'orange. What the company did is take muscle cells from the animals without hurting them and grew them in vats. The lab said that this is a process similar to brewing beer.

The company chose to work with chicken and duck because the two poultry meats are central to several cuisines around the world. The CEO of Memphis Meat, Uma Valenti is quoted as saying "This is a historic moment for the clean meat movement." The company believes that the way conventional poultry is raised generates too many problems for the environment, animal welfare and human health and is, in addition, inefficient. The cost of the meal was very expensive given the early stages of the process, around $9,000 for a pound of meat, but they are prioritizing the reduction of the cost as well as refining the taste, texture and nutritional value of the meat. The company has set a goal to serve this meat to customers by 2021.

Irving, Michael. Lab-grown chicken on the menu for the first time. March 15, 2017 [http://newatlas.com/lab-grown-chicken-memphis-meats/48434/]
http://newatlas.com/lab-grown-chicken-memphis-meats/48434/
http://www.usatoday.com/story/money/nation-now/2017/03/16/lab-grown-chicken-strips-made-animal-cells-debuted-startup/99259988/

Stronger than Steel

by Cybele Kappos

Two-dimensional graphene is an exciting material. It has unique electrical, thermal, optical, and chemical properties. In this normal form, graphene is only an atom thick and efforts to expand the material into three dimensions have so far failed, being a far weaker material than expected. In response to this problem, a team of MIT scientists have created a 3D version that has a sponge-like structure. The material is five times the density of steel but ten times as strong. Graphene could prove to have several practical purposes in engineering.

This team decided to focus on the geometrical configuration as opposed to the material itself. They initially analyzed the behavior of the atoms of the material and subsequently created a mathematical model to match the observations, then they used computer models to perform tensile and compression tests. By condensing small flakes of graphene under heat and pressure they created a stable porous structure that resembles coral. The structure is immensely strong and has a large surface area-to-volume ratio. This approach to graphene is similar to how paper can be folded into much stronger forms and can hold heavy loads.

Following the computer models, the team used a 3D printer to construct several configurations out of plastic. These were enlarged models of gyroids, which are porous structures. These shapes were tested in a similar way to the computer models and the results were compared.

The scientists found that the low density and strength of the material had more to do with the geometrical configuration than with the material. Tests conducted on other materials showed a similar gain in strength. This technology could be used to achieve stronger, lighter materials when created from polymers or structural concrete. Another suggestion for the porous material is for use in filtration systems for water or chemical plants.

Szondy, David. New 3D Graphene is Ten Times as Strong as Steel. January 9, 2017
http://newatlas.com/3d-graphene/47304/?li_source=LI&li_medium=default-widget

Non-toxic Thermoelectric Material

by Cybele Kappos

Thermoelectric devices are not new to the world of technology. Their function is to generate electricity through the thermoelectric effect, which converts differences in temperature into electric voltage. They are being used in a variety of ways including clothing and paint and are important because they reclaim some of the energy lost as heat in energy production. At the University of Utah, researchers have developed a new thermoelectric material that does not use the toxic materials that are found in conventional ones, and is still efficient and affordable.

Devices of its kind work by using a temperature difference between the two different sides of the materials. Charge carriers diffuse from the hot to the cold side and create an electric current. Typically, these devices use elements such as cadmium, telluride, and mercury which are toxic to humans as well as being expensive and inefficient. The researchers at the University of Utah are using calcium, cobalt, and terbium, thereby combatting the issues of safety and expenses. In addition, the elements are bio-friendly and eco-friendly.

The team already has plenty of suggestions for the use of the material. An example is jewelry that uses the material to power medical monitors by using the difference in temperature between body heat and air. Vehicles such as cars and planes could obtain extra energy in a similar way by harnessing the energy through the temperature difference between a cold exterior and a warm interior. Another interesting example is the suggestion of using the material to make pots and pans that could generate enough energy to charge a phone while cooking over heat.

The objective of the team is to use some of the energy that is wasted in typical energy processes due to heat. Power plants could use the material to produce more energy. According to a member of the team, around 60% of energy is wasted as heat. The first applications of the material will be in cars and biosensor.

Irving, Michael. Charging your phone with a cooking pot? New material could make it possible. March 20, 2017.
http://newatlas.com/thermoelectric-material-safe-inexpensive/48501/
https://phys.org/news/2017-03-lust-power-non-toxic-material-electricity.html

Sustainable Template For a Sustainable Villages

by Dena Kleemeier

In seeking biology research opportunities for the summer of 2017, I came upon a biology internship in a sustainable village in the heart of the Panamanian jungle. The name of the village is Kalu Yala (meaning sacred village), and markets itself as becoming the world's most sustainable residential community. The CEO Mark Stice intends on doing this by 'transcending real estate' and creating a "design to help improve historically impoverished regions in the tropics, and help to restore a global allocation of resources".

Kalu Yala was started in 2007 by an expat from Atlanta [Stice] who bought the land and started to develop it to be inhabitable and sustainable. The village uses only off-the-grid energy from 3 three primary sources: solar, micro hydroelectric, and an unspecified tertiary generator for any extra energy needs. The village also has a large aquifer naturally cleaning the water, as well as a rainwater catchment and storage for the dry season. Additionally, the village has a farm that they source most of their food from, to decrease the need for imported foods, and to strengthen the local economy. The housing in the village is designed and built by Studio Sky " a company focusing on smart dwellings for the tropics, and no more than 20 houses will be sold per year. As a result of these efforts, Kalu Yala is reported to be a carbon neutral (or negative) establishment.

Kalu Yala has an internship program, to which I applied. The program recruits interns for biology, sustainable agriculture, business, and wellness, and is in charge of helping to run many of the projects on the site. Though this is good practice for the interns themselves, they are not yet experts in their fields. Although there are professionals that help to run the village, it would seem that the efficiency and thoroughness of the project would not be to the same caliber if all of the projects were to be completed by professionals.

Petronzio, Matt. "Building the World's Most Sustainable Modern Town." Mashable. Mashable, 21 Feb. 2014. Web. 12 Feb. 2017.

The Most Beautiful Skyscrapers in the World
by Dena Kleemeier

Boeri Studio's completion of their vertical forest in Milan, Italy in 2014, catalyzed a series of similar buildings that are being built in cities across the world. Stephano Boeri Architetti was the architect behind the Bosco Vertical, in his effort to design buildings for European open areas that require regeneration. The Bosco Vertical is considered to be a 'vertical forest' and is a model for sustainable residential buildings, which is part of a larger project for metropolitan reforestation, which aims to improve urban biodiversity, and vertical densification of nature by having a living ecosystem in the middle of the city.

If these vertical forests were spread on flat land, they would individually cover 7,000 square meters of forest. The density and mass of forest "contributes to the construction of microclimate, produces humidity, absorbs CO_2 and dust, and mainly produces O_2". The forest naturally drastically increases the biodiversity in cities and has the potential to bring in birds and other animals. Architetti worked alongside botanists to choose trees with the appropriate heights for the project; were specifically grown for usage in the Bosco Vertical. Though the types of plants used on the building seems crucial the success of the building, it was not clarified which plants were used.

Boeri Studio's template has attracted significant attention over the past three years and has been utilized in many building templates, Stephano Boeri Architetti has personally traveled to polluted regions of Europe and China, to help develop projects that will help reduce CO_2 emissions, pollution, etc.

Nanjing, Sydney, Singapore, Belgium, Paris, Tokyo, Sao Paolo, amongst other cities, are actively working on, or have completed vertical forest projects as an anti sprawl device, to reduce pollution in urban environments, and to reduce energy consumption.

Homepage." Stefano Boeri Architetti. N.p., n.d. Web. 21 Feb. 2017.

Climate Change in National Parks
by Dena Kleemeier

Climate Central wrote a special report outlining the effect that climate change will have specifically on the National Parks. The author Brian Kahn, outlines how the park service must give up the notion that parks are natural places; "In reality they never were...parks have always been affected by how humans choose to manage them and our own definition of what natural is". Kahn expresses how we, members of society have a huge influence in the present and future conditions of the parks, however there is also some irreparable damage. The melted and melting glaciers in Glacier National Park are just one example of the casualties of the animals, plants, and parts of history that will irreversibly disappear. The article goes on to discuss the possibility of assisted migration to "lend a hand" to the wildlife that won't be able to migrate and adapt quickly enough. The author outlines an assisted migration project that the Park Service was currently undertaking, in which they brought bull trout to higher elevations to keep them away from non-native fish in their home waters. However, consideration of these kinds of projects has opened questions from researchers as to how proactive the park service should be in terms of the mobility of species. This kind of active role of scientists is controversial as it "puts scientists in the role of Noah, and gives them the jurisdiction as to 'who hops on the arc'". The article doesn't mention taking into account how assisted migration could potentially affect species and environment that live in the environments that the 'migrated' species move into.

National Parks, however are limited by financial resources that are allotted to them to help educate and preserve. They are actively working towards becoming more inclusive and accessible to expand their audience and to appeal to 'this generation'. In doing so, they hope that in 10 to 30 years they will receive the support they need. Additionally, they expressed how in the congressional environment the phrase "climate change" is likely to kill any funding requests, but they hope with a better understanding of climate change will justify asking for additional funding.

Kahn, Brian. "This is What The Future of National Parks Looks Like In the Face of Climate Change." National Parks | Climate Central Special Report. N.p., n.d. Web. 23 Apr. 2017.

New Techniques in Ultraviolet Light Use for Indoor Plant Growth

by Genevieve Kules

Since we were kids, we have been taught to cover our skin from ultraviolet (UV) radiation, but some types of UV rays have proven to be useful in situations like water purification and plant growth. Recent research has shown ultraviolet-C rays (the smallest, most energetic form) to be useful in indoor plant growing operations. While in the past artificial UV rays have been made using mercury, these are made using light emitting diode (LED) technology which is safer and more environmentally friendly than its predecessor.

UV light is classified into three different wavelengths: UVA encompasses wavelengths from 400 to 315 nanometers, UVB 315 nm to 280 nm, and UVC 280 nm to 100 nm. UVC is therefore the most energetic, fastest moving and dangerous type of UV ray. It is what kills bacteria in water and now is being used to regulate and enhance the growth of plants. When used in moderation, UVC light kills bacteria, prevents molds, and prompts plants to produce protein to protect themselves. These proteins can affect the smell, taste, and color of a plant and strengthen its defense mechanisms against diseases and viruses. UVC when used in short intervals, about fifteen minutes, can halt a plant's upward growth and redirect that growth to its branches. It can also cause it to produce more and healthier buds. But when used in excess, UVC light will burn and kill a plant.

One of the principal issues addressed in this study was efficiency of the LEDs. This was improved by using gallium nitride (GaN) instead of aluminum gallium nitride (AlGaN), and through polarization. This study recorded the lowest UV wavelength to date using GaN at 232 nm. The next step for this team of scientists is to package the energy as they would commercially and do a larger set of tests.

Cornell University. "Group blazes path to efficient, eco-friendly deep-ultraviolet LED." ScienceDaily. ScienceDaily, 3 March 2017. <www.sciencedaily.com/releases/2017/03/170303143124.htm>.

S. M. Islam, Kevin Lee, Jai Verma, Vladimir Protasenko, Sergei Rouvimov, Shyam Bharadwaj, Huili (Grace) Xing, Debdeep Jena. MBE-grown 232–270 nm deep-UV LEDs using monolayer thin binary GaN/AlN quantum heterostructures. Applied Physics Letters, 2017; 110 (4): 041108 DOI: 10.1063/1.4975068 <http://aip.scitation.org/doi/10.1063/1.4975068>

"Using Ultraviolet-C (UV-C) Irradiation on Greenhouse Ornamental Plants for Growth Regulation." StellarNet, Inc. N.p., 10 June 2016. Web. 05 Mar. 2017. <http://www.stellarnet.us/using-ultraviolet-c-uv-c-irradiation-greenhouse-ornamental-plants-growth-regulation/>.

"The Benefits of UV Light for Indoor Plant Growth." EYE Hortilux. Iwasaki Electric Co., 9 June 2016. Web. 05 Mar. 2017. <http://www.eyehortilux.com/blog/post/the-benefits-of-uv-light-for-indoor-plant-growth>.

New Measuring Tool Paves the Way Towards Fast, Low Energy Electronic Devices

by Genevieve Kules

New research in the field of quantum mechanics has found a way of combining spintronics and plasmonics in order to create rapid energy transmission in 1D systems. In simpler terms, this is the ability to measure and use together two different energy producing aspects of electrons, the spin waveforms and the charge waveforms, to eventually create high speed computers with low energy consumption.

The study done at the Tokyo Institute of Technology along with Nippon Telegraph and Telephone Corporation is the first to find a way of combining and measuring the two energies of electrons: the charge-density and spin-density of the waveforms. The researchers developed what they call a "spin-resolved oscilloscope" to measure the two energy forms and have them work together. Standard oscilloscopes measure either the spin signal or the charge signal or electronic computation. The charge signal of an electron is the sum of the spin-up and spin-down densities while the spin signal is the difference between the spin-up and spin-down densities. Until now, there was no way to measure both of these signals together.

This new technology leads the way towards faster, low energy data transfer. Whereas computers currently operate using binary code, quantum computing operates much faster and can compute far more complex equations than binary systems of standard computers we know today.

M. Hashisaka, N. Hiyama, T. Akiho, K. Muraki, T. Fujisawa. Waveform measurement of charge- and spin-density wavepackets in a chiral Tomonaga–Luttinger liquid. Nature Physics, 2017; DOI: 10.1038/nphys4062

Tokyo Institute of Technology. "Spin-resolved oscilloscope for charge and spin signals: Basic measuring instrument for future plasmonics and spintronics." ScienceDaily. ScienceDaily, 13 March 2017. <www.sciencedaily.com/releases/2017/03/170313135035.htm>.

Land Art Generator Initiative Promotes Collaboration between Art and Science

by Nina Lee

The Land Art Generator Initiative, founded in 2008, is a project that aims to create a more beautiful future for renewable energy technology. Taking into consideration technology's increasing integration into art, the initiative explores what art can do for technology. Thus far, the project has held competitions at three sites in the United Arab Emirates, New York City, Copenhagen, and Southern California. However, there are many components of the initiative. Their largest event is their biennial design competition, but they also hold invited competitions, commissions, requests for proposals, and facilitating participatory design processes within communities. All projects have a component of educational programming and community collaboration, as the initiative states on their website, "sustainability in communities is not only about resources, it is also about harmony."

The basic requirements for their projects are simple: capture energy from nature, convert it cleanly into electricity, and transform and transmit that electricity to a grid connection point. In addition, projects must respect the natural ecosystem of its intended site, and structures must be safe for the general public as well as for participants in any potential educational activities that may occur at the site. Once the proposal is approved, the Land Art Generator Initiative facilitates coordination between stakeholders, consultants, community groups, and local government.

When building sources of power in populated areas, creating visually appealing structures can add value to public space and spur local economies, therefore increasing livability for residents. Less quantifiable results could include inspiring onlookers to pursue further knowledge in any of the fields that contribute to the creation of a sustainably designed clean energy source. By utilizing the combined efforts of artists, architects, urban planners, environmentalists, scientists, and more, these projects can inspire innovation for more potential creators and also reach broader and more diverse audiences.

http://www.landartgenerator.org/project.html
http://e360.yale.edu/digest/public_art_or_renewable_energy_land_art_generator_initiative_
 santa_monica?title=public_art_or_renewable_energy_land_art_generator_initiative
 _santa_monica&id=4816

Introducing New Technologies to Hawaii's Growing Agriculture Industry

by Nina Lee

While farming and technology may seem antithetical to each other, Honolulu-based company Smart Yields believes in cooperation between the two. Smart Yields is a technology that can be used in mobile or desktop apps to provide and analyze data regarding air, soil, weather, and any other factors that could affect farming conditions. The app connects with a network of sensors and hardware to monitor the going-ons of the crops, as well as offer suggestions for improvement according to previous patterns and trends archived in various climate databases.

Smart Yields essentially applies the same data science used in fields such as finance and logistics to systems of farming. Especially regarding smaller farms on Hawaii, much of farming has been done based on traditional methods. However, with the ever increasing climate change, the farming environment is also changing, and thus requires a deeper understanding of the land. Smart Yields' goal is to use predictive analytics and provide more information to farmers who could use guidance during evolving farming conditions.

Due to the fact that Hawaii has historically housed large plantations that have since been shut down, the current agriculture industry is a little bit behind in comparison to United States mainland.

237

Smart Yields targets the needs of the growing independent agriculture industry that is progressively replacing the plantations. These small- to medium- sized farms that grow valuable crops such as cherry tomatoes, grapes, and cannabis are ideal for this type of technology since they require close monitoring. For example, information that would previously have been recorded using handheld oxygen meters could now be recorded digitally through Smart Yields' technology. The sensors can retrieve about 200 measurements every 24 hours.

http://www.hawaiibusiness.com/the-future-of-farming/
http://www.hawaiibusiness.com/data-driven-farming/
https://smartyields.com/about-us/

New "Sell By" Regulations take Aim at Food Waste

by Kieran McVeigh

In mid-February the Grocery Manufacturers Association and the Food Marketing Institute, announced that they are adopting new voluntary regulations for food safety labels. Currently there are as many as ten different labels, which food manufacturers put on their products ranging from "Sell By", to "Use By", to "Best if Used By. The new regulations replace these varied labels with just two, "Use By" and "Best if Used By". The "Use By" label will be used to label foods that are perishable, and indicate the date after which these foods are no longer safe to eat. The "Best if Used By" is for nonperishable foods and will display the date by which the manufacturer believes that the taste of the food will start to decline.

The current plethora of food safety labels causes confusion among consumers, who may believe labels are indicating the safety of the food, rather then best taste. This confusion causes undo food waste, which the new set of regulations seeks to reduce. Food waste, although not always brought in to the conversation surrounding climate change, is a significant contributor to greenhouse gas emissions. The Food and Agriculture Organization estimates that the greenhouse gas emissions (worldwide) for food waste amounts to 3.3 Gigatons of CO_2. Comparatively, if this were a country's emissions it would be the world's third largest emitter of greenhouse gases. Forty two percent of US greenhouse gas emissions come from the food production cycle, and as much as 40% of that food in the US wasted, this is an area where there could potentially be large reductions in greenhouse gas emissions.

Although these guidelines are voluntary, large retailers such as Walmart have pledged support, which suggests they may be widely implemented. That these guidelines originate from the Grocery Manufacturers Association and the Food Marketing Institute increases the likelihood companies will comply, as representatives from many large companies are members of these organizations. These guidelines represent an important step to reducing the emissions due to food waste yet there is still much more to be done target food waste at all steps of the food production and consumption cycle.

Dewey, Caitlin. "You're about to see a big change to the sell-by dates on food." The Washington Post. WP Company, 16 Feb. 2017. Web. 27 Feb. 2017. <https://www.washingtonpost.com/news/wonk/wp/2017/02/16/a-barely-noticeable-change-to-how-food-is-labeled-could-save-americans-millions/?utm_term=.b77671e1ceab>.

"Food Waste Worsens Greenhouse Gas Emissions: FAO." Climate Central. N.p., 22 Sept. 2013. Web. 27 Feb. 2017. <http://www.climatecentral.org/news/food-waste-worsens-greenhouse-gas-emissions-fao-16498>.

Gunders, Dana. "Wasted: How America is Losing Up to 40% of its Food." Ndrc.org. National Resources Defense Council, Aug. 2012. Web. 27 Feb. 2017. <https://www.nrdc.org/sites/default/files/wasted-food-IP.pdf>.

No More Floods

by Byron R. Núñez

Flash flooding causes billions of dollars in property damage every year, with more than two-thirds of the damage being strictly attributed to water running off pavement. Tarmac, a United Kingdom company, has created a new type of porous concrete that can absorb up to 4,000 liters of water per square meter. The company hopes that its concrete can end flash flooding in urban areas once and for all.

Topmix Permeable concrete works by having different layers, which are designed to drain water as quickly as possible. The first layer is made up of large pebbles through which water can drain almost instantly. This layer is followed by an "attenuation layer", which feeds the water into a drainage system that is connected to the city's groundwater reservoirs. The water from these reservoirs is then used for irrigation, drinking water, swimming pools, and firefighting purposes. Permeable concrete has been around for the past 50 years; it is mainly used under pavement to help with drainage. Tarmac researchers engineered a way to make a surface layer version that can withstand the weight of heavy traffic.

Traditional pavement is also dangerously hot during summer months and creates precarious situations for animals walking barefoot. A benefit of Topmix permeable concrete is that it is cooler than traditional asphalt. During periods of rising temperatures and intense rainfall, water stored within the system evaporates creating a cooling effect that reduces surface temperatures. If used on sidewalks, this concrete could protect dogs, and other animals, from getting their paws burned by the hot concrete. However, in areas that suffer from extreme cold temperatures, the concrete is likely to be damaged since freezing water that can expands and cause the surface above to buckle and crack.

Daily Mail
http://www.dailymail.co.uk/sciencetech/article-3243247/An-end-puddles-Bizarre-thirsty-concrete-sucks-hundreds-gallons-water-minute.html

Urine-Powered Generator

by Byron R. Núñez

Four Nigeria teenage girls have figured out a way to use a liter of urine as fuel to power a house for six hours. The urine-powered generator uses electrolysis to isolate hydrogen gas from urine which is then used to power a generator. The system has four steps, first, urine is put into an electrolytic cell, which separates out the hydrogen, second, the hydrogen enters a water filter for purification, which then gets pushed into a gas cylinder, third, the gas cylinder pushes hydrogen into a cylinder of liquid borax, which is used to remove the moisture from the hydrogen, fourth, this purified hydrogen gas is pushed into the generator. It currently takes more energy to extract hydrogen from urine than what is eventually produced by the generator. However, the young Nigerian scientists are working with researchers to make the generator more efficient.

The idea of using urine as fuel is not something new. Gerardine Botte, a professor of chemical and biomolecular engineering at The Ohio University, has been working on practical ways to make urine a more useful hydrogen source. Botte explains that current power generators consume a lot more energy than they produce and that the energy equation gets even more skewed by the inefficiency of the generator used. While the generator used by the Nigerian scientists is not as efficient as it could be, the young girls have proven that their innovation can one day solve issues of energy accessibility that plague many developing nations.

In Lagos, Nigeria, for example, power outages happen multiple times a day and only those who can afford an expensive backup generator have uninterrupted access to electricity. This leaves many residents without reliable access to the grid. The urine-powered generator created by the Nigerian girls might have no net gain in power, but it is a step in the right direction. It is also important to note that a breakthrough for this innovation is the fact that it does not release carbon monoxide, which means it can be used to fuel Nigerian homes without filling them with toxic gasses. The system also does not pose the risk of explosion since the inventors used one-way valves throughout the device as a safety measure.

As this technology continues to evolve, it could one day be used to power vehicles since gasoline-powered internal combustion engines can be easily converted to run on hydrogen. This raises the question of whether there is potential for pee-powered cars in the future.

The Guardian
https://www.theguardian.com/powershop-powering-better-
        future/video/2016/nov/30/pee-powered-generator

ENERGY EFFICIENCY

Energy Efficiency

Wireless Power

by Dominique Curtis

Disney Researchers have developed a new method for wirelessly transmitting power throughout a room that enables people in that room to charge their electronic devices, potentially eliminating the need for electrical cords or charging units. The researchers used a method called quasistatic cavity resonance in a 16' by 16' room in their lab to generate magnetic waves that filled the entire room and made it possible to charge several cellphones, fans, and lights simultaneously (Science Daily, 2017).

The quasistatic cavity resonance (QSCR) method is accomplished by first inducing electrical currents in the metalized walls, floor, and ceiling of the room, then generating uniform magnetic fields that permeate the rooms interior. This allows power to be transmitted efficiently to receiving coils that operate at the same resonant frequency as the magnetic fields. The Disney researchers' room-scale wireless power demonstration used a 16' by 16' foot room with aluminium walls, ceiling, and floor bolted to an aluminium frame. A copper pole was put in the center of the room, and a small gap was put in the pole to insert discrete capacitors.

While delivering useful amounts of power in a innovative way this method must also be safe for the people who will be in these spaces. So there are a few safety concerns that the researchers are addressing. The results from the safety simulation research show that it is possible to safely transmit 1.9 kilowatts of power to a receiver at 90% efficiency, the equivalent of charging 320 USB powered devices. (Chabalko, 2017)

Moving forward, Disney researcher Alanso Sample predicts that in the future "it may be possible for electrical power to become as ubiquitous as WIFI" (Science Daily, 2017). There is also talk about future plans to retrofit existing structures with modular panels or conductive paint. Researchers also say it's all about scale. They see no difference in scaling down to a cabinet or storage box size or scaling up to a warehouse size building using the QSCR method. Soon we could be charging our phones at power hot spots.

Science Daily. "Wireless Power Transmission Safely Charges Devices Anywhere within a Room." ScienceDaily. ScienceDaily, 16 Feb. 2017. Web. 28 Feb. 2017.

Matthew J. Chabalko, Mohsen Shahmohammadi, Alanson P. Sample. Quasistatic Cavity Resonance for Ubiquitous Wireless Power Transfer. PLOS ONE, 2017; 12 (2): e0169045 DOI: 10.1371/journal.pone.0169045

New LED Outdoor Lights Pose More Negative Effects than Anticipated

by Genna Gores

Over the past few years many U.S. cities changed their outdoor lights from the ordinary sodium-vapor streetlights to energy-efficient LED lights. In the United States street lights make up 30% of the energy used for outdoor lighting, so switching over to LED lights could lead to the U.S. saving 662 trillion British thermal units of energy—the same amount of energy needed for 5.8 million U.S. homes for a single year. The new lights could also save millions of dollars in energy spending; for example, Newton, Massachusetts, could save $3 million and prevent 1,240 tons of carbon dioxide emissions with a population of just 80,000 people. It is clear that this switch is beneficial for many reasons, but it comes with side effects.

Many people dislike the bluish tint emitted from the LED lights, and complain that the light pollution in their towns is worse. Jeff Hecht, writer for IEEE Spectrum, explains that, "The bluish LEDs were a stark counterpart to the orangish high-pressure sodium lights that came before them. Switching from the warm sodium lights to those LEDs was like going from a subtropical sunset to high noon at the equator." Beyond complaints from citizens there is also new research coming out that the blue light affects sleep patterns of humans and also the nighttime activity of nocturnal animals. According to this research, the blue LED lights affects one's circadian sleep rhythm five times more than conventional street lights, and also affect sleep patterns in the same way as looking at a smartphone or a computer screen before going to bed. Despite impacting humans, the lights also have harmful effects for nocturnal animals such as bats and moths. Baby sea turtles, which are already endangered, suffer from the blue LED lighting at coastal resorts. Because sea turtles are guided to the ocean by the moonlight, these lights confuse the turtles and strand them on the beach and leaving them vulnerable to predators.

While there are ways to mitigate the harshness of blue lights it could potentially be costly to local governments. By changing the light bulbs to LED versions that emit a more yellowish light—by using compounds called phosphors that absorb blue light and then emit yellow light—the municipalities would lose efficiency and savings. The conversion of blue to yellow light causes an energy loss making light bulbs less efficient, and since 10% of U.S. has already switched to these bluer versions it could cost billions of dollars to replace. So, to cut down costs it is imperative to stay on the blue side of the spectrum. Towns that have already installed the LED street lights are now looking into wireless control systems rather than replacement. The hope is that they could remotely dim lights when complaints arise and even dim the lights in the latest hours of the night. LED lights are the future of outdoor lighting, and it is now up to municipalities to find the balance between savings and a pleasant environment; until the industry figures out how to produce equally effective ones with a warmer color temperature.

[http://spectrum.ieee.org/green-tech/conservation/led-streetlights-are-giving-neighborhoods-the-blues]

[http://spectrum.ieee.org/green-tech/conservation/led-streetlights-are-giving-neighborhoods-the-blues]

[http://spectrum.ieee.org/green-tech/conservation/led-streetlights-are-giving-neighborhoods-the-blues]

Chicago Switches Government Buildings to Renewable Energy

by Genna Gores

Mayor Rahm Emanuel of Chicago, Illinois, announced that the city of Chicago will convert 900 of their government buildings' electricity use to renewable energy sources by 2025. These buildings will include publics schools, city colleges, the Chicago Park District fieldhouses, administrative buildings, and the Chicago Housing Authority. As of now, these buildings use 8% of all the electricity in Chicago, a significant percentage, and this amounts to 1.8 billion kilowatt hours. According to the Chicago Sun Times, this amount of energy can power 295,000 Chicago homes, and would take 300 wind turbines. The City of Chicago will accomplish this goal by utilizing power purchase agreements with offsite energy producers of solar and wind energy, installing solar panels and/or windmills on city buildings or public land, and buying energy from the Illinois Renewable Portfolio Standard. Currently, Emanuel has not released a statement on how much the city is willing to spend, but as of now they have determined less than a 1% net increase in total energy costs in the long term. They believe that the long-term savings from using renewable energy will outweigh the upfront costs of the power purchase agreements and solar installations. Chicago has already installed solar panels on the roofs of Richard J. Daley College and the Dawson Technical Institute, and this change to renewables already amounted to $16,000 in savings on energy. In a time when the federal government is pulling funding for renewables, and saying no to clean energy, it is up to cities to make the change. Chicago's commitment to renewable energy demonstrates that with power purchase agreements and adding their own electricity to the grid that switching to renewable energy is feasible.

[http://chicago.suntimes.com/politics/900-chicago-government-buildings-to-switch-to-renewable-energy/]

Cold Sintering: An Energy-Efficient Method of Firing Ceramics

by Nina Lee

Ceramics make up many of the literal building blocks of our society: cement, brick, porcelain, and tile. In order to give the material the durability it is utilized for is the process of firing clay, traditionally done at temperatures over 1,400°C. Powering the firing process requires an immense amount of energy, so researchers at ETH Zurich have come up with a method of firing that can occur at room temperature. The alternative firing process, called cold sintering, is modeled after the process of rock formation found in nature where pressure is used to compress calcium carbonate nanopowder and water into limestone. The resulting material is just as strong as concrete, if not stronger.

Because the calcium carbonate nanopowder contains particles of such a fine size, the compacting process takes only one hour for a product about the size of a one franc piece. This is comparable to the typical ceramic manufacturing time. Anything larger then that would require more pressure that the average hydraulic press can provide. The trajectory of this project would most likely be focused on creating small pieces such as bathroom tiles.

Another added benefit to cold sintering is that, according to researchers, carbonate nanoparticles could potentially be produced using CO_2 captured from the atmosphere or by products of thermal power stations. This would be accomplished by allowing the captured CO_2 to react with powdered rock to produce the carbonate used in the cold sintering process. This would allow for a more sustainable alternative to cement production, which requires large amounts of both energy and CO_2.

https://www.sciencedaily.com/releases/2017/02/170228084222.htm
http://www.nature.com/articles/ncomms14655

OhmConnect Pays You to Save Energy

by Kieran McVeigh

Often, reducing energy consumption and saving money go hand in hand. OhmConnect a company based in Southern California is strengthening this bond. OhmConnect offers a free service that pays people to reduce their power consumption, allowing users to cash out their energy savings at the end of the year through Paypal. OhmConnect monitors the California ISO's energy forecasts and current energy consumption. When actual energy usage is outstripping predicted energy usage, OhmConnect sends out alerts to their users notifying them that they are in peak energy usage, then if users reduce their power usage below their predicted energy usage during this period they receive credits from OhmConnect.

OhmConnect makes their money by offering contracts to electric companies promising they will create energy savings, theoretically preventing the energy companies from having to turn on the more expensive backup generators, and saving everyone money in the process.

Currently over 100,000 Californians use OhmConnect, with the average user saving between $50 to $150 dollars a year. Although OhmConnect is only available in California, as more states adopt regulations that allow OhmConnect to function they plan on expanding. Another potential boon to OhmConnect is that with certain "smart" devices, like smart thermostats, or even Teslas this energy reduction can happen automatically, so whenever energy usage surges, your Tesla stops charging for a few hours or your thermostat turns off the AC for a few hours. As more and more devices run on some sort of smart network, this could make OhmConnect even easier to use and more effective.

OhmConnect has also recently harnessed the power of gamifying their product assigning ranks to the customers, so customers are able to see how they compare against others users. Customers can also participate in energy saving teams where their savings can be donated to the organization of their choice, making OhmConnect not just a personal money saver, but a social activity as well.

Lobet, I. (2017, March 07). Energy Savings Can Be Fun, But No Need To Turn Off All The Lights. Retrieved March 20, 2017, From http://www.npr.org/sections/alltechconsidered/2017/03/07/518175670/energy-savings-can-be-fun-but-no-need-to-turn-off-   all-the-lights

Finley, K. (2015, February 15). The Internet of Anything: The System That Pays You to Use Less Electricity. Retrieved March 20, 2017, from https://www.wired.com/2015/02/ohmconnect/

San Diego Installs Smart Sensors

by Kieran McVeigh

The Internet of Things (IOT) refers to the ever growing range of devices that are connected to the internet allowing increased connectivity of "things" that have never been connected to a network before. San Diego is taking this trend very seriously, announcing in early 2017 that they will install 3200 smart sensors throughout the down town area, in conjunction with replacing 14,000 streetlights with new LED streetlights.

These upgrades are predicted to save San Diego as much as 60% of energy used annually to light its streets, in part due to the increased efficiency of the LED light bulbs as well as the interconnectivity of the new smart sensors. The smart sensors will have a variety of abilities including monitoring traffic, carbon emissions, and detecting where gunshots are fired. One of the first initiatives that these smart sensors will be used for is to monitor the need for lighting the streets, so lights can be dimmed when there is less need, decreasing light pollution throughout the city. This capability will not only save the city energy but will help increase the lifetime of the new lights as they will not be used as often.

While the city's initiatives to save energy via more efficient lighting is one exciting part about the installation of these sensors, another important part is that data from the sensors will be made public, and available to third party developers. This open data will allow developers to drive innovation within the city from creating apps to find parking to who knows what else. By making these data available the city opens up a broad range of opportunities and possibilities. It will be exciting to see what happens with all of these new data about how the city works and see what is developed. It seems the sky is the only limit.

Woods, E. (2017, April 21). San Diego Aims to Set the Pace for Smart City Networks. Retrieved April 24, 2017, from https://www.forbes.com/sites/pikeresearch/2017/04/21/san-diego-aims-to-set-the-pace-for-smart-city-networks/#204af9e2756b

San Diego to Deploy Words Largest "Internet of Things" platform. (2017, February 21). Retrieved April 24, 2017, from https://www.sandiego.gov/mayor/news/releases/san-   diego-deploy-world%E2%80%99s-largest-city-based-%E2%80%98internet-things%E2%80%99-platform-using-smart

America's Most Energy-Intensive Crop

by Byron R. Núñez

In the United States, the legalization of medical and/or recreational marijuana has increased the demand for the commercial cultivation of the plant, which happens to be America's most energy-intensive crop. On average, it takes 13,000 kilowatts annually to grow five pounds of marijuana, which is more electricity than the average United States' household uses in an entire year. This results in 1% of the entire nation's energy output going towards the cultivation of the crop. Furthermore, every plant requires about six gallons of water a day to thrive. Cannabis may grow green and make green but growing it is far from green.

In Maine, producers are making efforts to improve inefficient growing methods. However, they note that marijuana is almost entirely grown indoors since the state's seasons limit outdoor growth to only a few months. Moreover, marijuana remains a schedule I drug under the United States' Controlled Substance Act (USCSA), which means that transportation of the crop from one state to another violates federal law. This classification prohibits states like Maine from importing marijuana from other states that can grow the crop more efficiently.

Many growers are also ineligible for federal grants designed to help farmers become more sustainable. The grants would allow them invest in practices that are less energy and water intensive. With the legalization of marijuana in California, the country's most populous state, more sustainable cultivation methods are needed. However, if the federal government continues to list marijuana as a schedule I drug, the possibility of making this crop more sustainable is slim.

Portland Press Herald

http://www.pressherald.com/2017/01/15/whats-the-most-energy-intensive-crop-in-america/

A Round the World Trip with Zero Emissions: The Energy Observer

by Nadja Redmond

The Energy Observer, a former multi-hull race boat that has been converted into an energy conscious, solar paneled vessel, is ready to set sail. Pulling its power from wind turbines and self-generated hydrogen fuel, the boat will embark around the world voyage this spring, starting with Paris and returning to Europe in the spring of 2023 after visiting 50 countries along the way. Through those six years, the boat will make 101 stops around the globe. The boat, which cost about $5.25 million to update with renewable energy in mind, was first built in 1983 and had a successful career racing in open ocean sailing competitions. The boat was brought out of retirement in 2015 by skippers Frederick Dahirel and Victorien Erussard, who conceived of their idea with the future in mind. French Environmentalist Nicolas Hulot believes the boat demonstrates "that there are many solutions for energetic transition," that "all solutions are within nature."

Drawing on solar and wind power primarily, and hydrogen fuel generated from electrolysis of sea water when those resources are lacking, the boat presents a fantastic peek into what green, energy-conscious travel may look like in coming years. The diverse balance between the renewable energies and the hydrogen production and storage is most likely the model that will be developed by these initiatives, according to Florence Lambert and CEA Liten, the French institute that developed the boats fuel and energy systems. This boat taking a multinational trip around the globe is significant. This ability for renewable, potentially cost effective energy to power such a trip hopefully prefaces further strides in energy conscious transportation initiatives, and signifies that smaller feats will be even more achievable soon.

[http://www.seattletimes.com/business/zero-emission-boat-prepares-for-round-the-world-odyssey/].

[http://www.sltrib.com/home/4807540-155/zero-emission-boat-prepares-for-odyssey].
http://liten.cea.fr/cea-tech/liten

The Future of Lab Grown Meat

by Nadja Redmond

In the future, it is very likely meat could be grown from single cells at home using personal meat-making machines. Scientists are currently developing methods to produce animal-free beef, chicken, turkey, and fish. Last month, San Francisco-based startup Memphis Meats revealed lab grown fried chicken. Taste testers described a product that was basically chicken with a slightly spongier texture than normal.

The logic behind lab grown meat stems from the idea that naturally in animals, muscle tissue—meat—grows from just a few cells into a larger section. Scientists concluded that this process should be able to be replicated in a lab using self-renewing cells taken from the appropriate animals, and many hope for their lab grown meat to be ready for mass production soon. Because these products would reduce environmental cost, increase health benefits, and protect animal welfare, this process could change the meat industry drastically in the future.

Over the past half-century, meat consumption has doubled because of population growth and changing trends; by 2050, meat production is estimated to increase from 259 to 455 million tons each year. Today, meat (and dairy) production accounts for 70% of global water consumption, 38% of land use, and 19% of world's greenhouse gas emissions. While lab-grown meat would help reduce these percentages, the process comes with its own set of energy consumption issues, with the electricity and heat needed to power the labs. Also, another challenge would be producing enough meat for the cost to be reduced to that of meat already harvested from animals. As commercial interest in lab-grown meat increases, other scientific breakthroughs around stem cells or tissue engineering will hopefully make way for the appropriate changes for lab-grown meat to make a significant change in the meat industry

[http://www.nbcnews.com/mach/innovation/lab-grown-meat-may-save-lot-more-farm-animals-lives-n743091].

The Tesla of Architecture: Freiburg's Smart Green Tower

by Yerika Reyes

As a city committed to achieving carbon neutrality by 2050, Freiburg, Germany is in the process of building the Smart Green Tower, a building that will power itself and the surrounding area. The architect, Wolfgang Frey, from Frey Architekten, took the concept of a Tesla cars that run on lithium-ion batteries and transferred it to buildings. Frey has designed a building in partnerships with Siemens AG and the Fraunhofer Institute for Solar Energy Systems (ISE) that is entirely solar powered. The Smart Green Tower will compromise of a main building with two wings with a total gross floor area of approximately 15,000 square meters, 70 one-to-four room apartments and office space. The building has many facets that will help to make it energy efficient. The inside of the apartments will all have sensors that will be programmed to self-adjust the heating when a window is opened or closed, and conserve energy by switching off a light when no one is in the room. These innovations in the Smart Green Tower are powered by a building management system (PowerManager) from Siemens will use a direct current (DC) intermediate circuit rather than the standard alternating current (AC) transmission and distribution of electrical energy which has a high efficiency loss. This will allow direct management of electricity production, consumption, and storage. Energy will be stored in a lithium-ion battery, enhanced by vanadium redox flow batteries this will permit the building to connect to the rest of the district. Additionally, the building has high performance photovoltaic (PV) panels which have excellent low-light performance. These tools have made the design of the Smart Green Tower not only sustainable in isolation, but helping the entire community.

http://www.freyarchitekten.com/en/projects/smart-green-tower/
https://www.siemens.com/customer-magazine/en/home/buildings/light-and-building/freiburgs-smart-green-tower-technology-meets-tradition.html

The Alternative to Styrofoam
by Yerika Reyes

A Swedish team has developed a material similar to Styrofoam called Cellufoam made from biodegradable wood-based material. Styrofoam by a Dow Chemical Co. trademarked form of polystyrene foam insulation, introduced in the U.S. in 1954. Styrofoam is 98% air, making it lightweight and buoyant. It absorbs shocks and insulates, and has uses from the to-go coffee cups to a sturdy bicycle helmet to coolers, or cushioning material in packaging. However, it is petroleum-based, and it's a synthetic material that takes at least 500 years to decompose. It has made a sizable contribution to landfills across the world for decades. In 2015 there have been some worms have been cultivated to eat foamed polystyrene, according to recent studies but the synthetic material remains the primary source of litter in U.S. cities – and is the most prevalent pollutant in American waterways, according to research conducted by Washington University in St. Louis.

Cellufoam was developed by Wagberg, along with colleagues now at Stockholm University and the Wallenberg Wood Science Center, who make up the new startup called Cellutech. The material is made with tiny wood cellulose fibers, which are blended with a foaming mix of water and air. The process of Pickering emulsion, which incorporate solids into the mix to prevent it from separating and coalescing, stabilizes the air bubbles in the material. The ample sources of wood in Sweden, where they regularly plant more trees than are harvested, could make the material even more vital.

Cellutech teamed up with designers Ramsus Malbert from Design Studio Materialist, and Jesper Jonsson to construct a bicycle hamlet as example of one use. However, the Cellufoam material could be employed in uses beyond the bicycle helmet, including potential flame retardants, water filtration and other ways. The properties of the material are suitable for use in lightweight construction, packaging materials and absorbent products.

http://www.cellutech.se/
https://phys.org/news/2016-01-swedish-scientists-wood-biodegradable-renewable.html

https://www.kth.se/en/aktuellt/nyheter/a-renewable-and-biodegradable-alternative-to-
    styrofoam-1.618724

Better Energy Programs for Chicago's Underserved Communities

by Yerika Reyes

One of the largest polluters in Chicago is its buildings. While energy efficiency is considered when architects and developers plan to erect new buildings, most Chicagoans live in older homes that are energy -intensive. A study from Elevate Energy reveals Chicago's built environment and the residents that occupy the city's housing stock. The study is a part of the Energy Efficiency for All Initiative (EEFA), and reveals that three out of four Chicago residents live in older brick and frame multi-family buildings that were constructed before 1942. While this may not be unexpected to many residents, it's critical to consider how much energy it takes to heat and cool these buildings and the costs to supply this energy. According to the study, Chicago is home to about 1 million units of housing in multi-family buildings. About 75 percent of these units were constructed before 1942. Elevate Energy estimates that of these units roughly a third are condominiums, while the remaining two-thirds are rental apartments.

Condo associations and flippers will often modernize much of the housing stock in more affluent areas, however in lower income neighborhoods, many buildings remain energy inefficient. The causes of inefficacy include poor insulation, old windows, and leaky pipes causing older homes to use much more gas, electricity, and water than newer buildings. These inadequacies lead to both steeper utility bills and cause pollution and greenhouse emissions. Based on their findings, Elevate Energy suggests that better energy programs should be offered for Chicago's underserved neighborhoods. Based on their housing stock and number of lower income families living in multi-unit buildings, the report highlights six communities would benefit from better energy efficiency programs: Auburn-Gresham, Austin, Humboldt Park, the Lower West Side, North Lawndale and South Lawndale. According to Elevate Energy, by targeting these six neighborhoods, up to 80,000 units of housing in 25,000 buildings stand to benefit from efficiency programs.

Three years ago, the mayors of several major cities across the country joined together under the City Energy Project to focus on ways to reduce greenhouse emissions from buildings. However, the focus has been on the most visible offenders which include high-rise office, hotel, and residential buildings that blanket the downtown area. Nonetheless, improving the energy efficiency of Chicago's older housing stock can only improve the city's greenhouse emissions and lower costs to property owners.

http://www.elevateenergy.org/segmentation-chicagos-multifamily-market-improve-
    program-design/
http://energyefficiencyforall.org/

http://www.cityenergyproject.org/

Researching the Fashion Industry and its Effects on the Environment

by Yerika Reyes

When looking for culprits for the cause of climate change we often quickly go to oil companies, large coal-fired power plants, inefficient cars, and massive freighter moving our online purchases around the globe. However, an industry not heavily researched but sometimes used as an example of heavy inefficiency is the fashion industry. Although there have been news sites and blogs like Ecowatch and Forbes that have claimed that the "fashion industry is the second largest polluting industry in the world," there is not much data to substantiate this claim. Instinctively we can agree with such a claim because if we think of the supply chain of fashion world, that does not seem a far-fetched declaration. Furthermore, when we look at the materials that are used, we do know that an estimated 50 million tons of polyester—a petroleum product—were produced in 2015. We also know that growing cotton, particularly when pesticides, herbicides, and oil-powered machinery are used, is also a large carbon emitter. One garment is composed of many materials that get moved around the world in oil-guzzling ships. The material is spun in one country, sewn in another factory, buttons and zippers are obtained from another factory, and then finished in yet another factory. These factories may be powered by coal and generators are just one of many stops along the way to being finished. Finally, a garment gets packaged and shipped to another continent, bought, worn for a short time, and then either gets tossed in a landfill that emits greenhouse gas and methane, or it reenters the garment supply chain in secondhand markets.

Although we can think through the ways in which the fashion industry is harmful, when looking for sources there is no official report that can confirm that the fashion industry is energy inefficient and therefore harmful to the earth. Many sustainable fashion experts cite the Danish Fashion Institute. "We don't believe the statement to be accurate either, but we are aware that it has become a popular misconception," the Danish Fashion Institute said in an email to reporter Alden Wicker. "We can, however, tell you that fashion is one of the most resource-intensive industries in the world, both in terms of natural resources and human resources." The Danish Fashion Institute is working on a report that will come out in May to make the impact of the fashion industry clear. It is true that fashion impacts a large range of the environment. It impacts agriculture (cotton, flax, hemp), animal agriculture (leather, fur, wool, cashmere), petroleum (polyester and other synthetics), forestry (rayon), mining (metal and stones), construction (retail stores), shipping, and, manufacturing. The complex and multilayered supply chain provides both a challenge and an opportunity for climate advocates. It is imperative that research be done in this field to gain a more accurate image of the different ways in which the fashion industry is harming the earth. Additionally, we can start working towards eco-friendly solutions.

http://www.ecowatch.com/fast-fashion-is-the-second-dirtiest-industry-in-the-world-next-to-big--1882083445.html

https://www.sei-international.org/publications?pid=1694

https://www.bsr.org/files/bsr-report/bsr_report_2009.pdf

https://www.forbes.com/sites/jamesconca/2015/12/03/making-climate-change-fashionable-the-garment-industry-takes-on-global-warming/#60a344cc79e4

New Way of Recycling Fast Fashion

by Yerika Reyes

We own a lot of clothes. A 2014 Alliance Data survey found that 37 percent of women own between 25 and 49 blouses. An archived survey from Design for Living Magazine held by the New York Public Library showed that women used to own anywhere from three to eleven blouses at most. This increase in shopping habits translates to an increase in waste, especially when it comes to fast fashion. Fast fashion is created by retailers like Forever21, H&M, and Zara who purposefully design their clothes to be worn only a handful of times and are made from textiles that are difficult to recycle. In the United States in 1960, the average citizen produced twenty pounds of textile waste per year. In 2010, that number rose to 85 pounds per person, according to data from the Environmental Protection Agency.

Thus current fabric recycling process is not as sustainable as one might think, to appreciate the new recycling method that Finnish scientists from Alta University presented at the American Chemical Society this year. Most clothing isn't recycled; it is reused. Portions of it end up at thrift stores and second-hand shops, but the majority of our old clothes are shipped to countries in Asia and Africa, where development experts argue their influx hurts local textile markets. Another small amount of clothes is cut down to make industrial cleaning rags and insulation, while an even smaller percentage is turned back into raw fiber elements, a process known as fiber conversion. Fabric conversion involves physically "tear[ing] fabric apart to maybe make some new yarn by adding a bit of new fibers," said Simone Haslinger, a researcher in the department of bioproducts and biosystems engineering at Alta University. "So, they get this blend of new and old fibers." After fabric is put through an industrial shredder, the resulting textile is not as strong or as versatile as its source material. Fiber conversation often results in down cycling into carpet or paper instead of being recycled into a new shirt. This current method of recycling helps to reduce some waste, however it is a completely closed system. Haslinger and her colleagues at Alta University have been working on attempts to close the loop and turn an old shirt into a fiber than can be spun into a new shirt. Five years ago, a team of researchers lead by Herbert Sixta, developed a technique that dissolves cellulose (the plant fiber in fabrics like cotton, rayon, model, and tencel) and turns it into a material that is stronger than rayon. However, initially this technique only worked on entirely natural materials—not the poly-fiber blends that most modern clothing is made from. This new technique, developed by Haslinger, extracts the natural fibers through a process they call Ioncell while leaving the polyester behind.

To transform a poly-cotton blend t-shirt into a new shirt, Haslinger exposes the fabric to an ionic solvent—a salt—that's chemically modified to have a lower melting point. "When you compare it to sodium chloride, for example, it has a melting point of around 1,400F," said Haslinger. "By chemically modifying this component, we could get a melting point that's around 140F, but it still has a lot of properties that are similar to salt. That's why it's actually able to dissolve cellulose, which is hard to dissolve otherwise." Exposing the material to a solvent extracts the cellulose while exposing it to an anti-solvent creates a gooey mass of cellulose. The mass is put through a press that extracts thin strands of fibers which can then be woven into new fabric as strong as tencel, a silky fabric made from wood pulp that's beloved by some environmentalists because of its closed-loop process. The whole process may seem rather labor intensive, however the chemical input it takes to plant, grow, harvest, spin cotton into threads, and then weave it into fabric is more laborious. Nevertheless, the Ioncell-F technique is not perfect. It cannot handle all fabrics, for example elastane, also known as spandex, which make it difficult for fast fashion shops to move into this new recycling method. Another challenge that this process has, is that it requires that the fabrics be bleached and then dyed again after the fabric is remade. The researchers hope that with better industry cooperation, they'll be able to sort fabrics by color in the future. If all blue textiles showed up at the recycling plant in the same pile, the Ioncell-F process could keep their dyes intact thus reducing the environmental impact of the new fabric even further.

http://puu.aalto.fi/en/research/research_groups/biorefineries/ioncell_f/
https://www.acs.org/content/acs/en/pressroom/newsreleases/2017/april/upcy cling-fast-fashion-to-reduce-waste-and-pollution.html

Smart Tech and the Military

by Yerika Reyes

As technology and the social landscape change so must the way our military reacts. The Pentagon recently directed all its bases to raise their threat level, because of a social media threat. Accounts connected to ISI had posted the names and home addresses of US military personnel with a directive to terrorists to attack. Military bases across the country began implanting stricter security. The security threat is one way in which military bases are always responding to new technologies to make stronger and more comprehensive security responses. Smart technologies are proven and offer significant advantages for military installations. The smart military base is overdue and could improve many aspects of military bases, including security.

A smart base must employ: artificial intelligence, the Internet of Things, machine automation and robotics, and data analysis, (to name a few) to improve the quality and speed of its functions and services. Taken together, they are able to collect and process large amounts of data that facilitate more economical operations and help military staffers make better decisions. Employing smart technologies to military bases has already started. Army and Navy bases in Georgia recently opened smart energy solar plants that allow them to operate independent of the local power grid, a critical capability in an emergency. Fort Bragg is experimenting with driverless vehicles to transport wounded soldiers across base to rehab appointments.

The Department of Defense (DoD) oversees thousands of buildings across more than 5,000 locations on more than 30 million acres of land. The Government Accountability Office reports that DoD installations have a replacement value of $880 billion. Large efficiencies can be realized from using technology to help monitor and manage power, water, and construction costs. The Army's smart energy program has already reduced costs by nearly $150 million. They Army leases unused base real estate and returns money to taxpayers while also serving the community around the base. A geothermal power plant at the Naval Air Weapons Station in China Lake, contributes nearly 1.5 million megawatts-hours of electricity to the grid each year. To gain all the benefits of smart technological bases we must work on technologies that facilitate beneficiary relationships with the surrounding community and the personnel at the base.

http://thehill.com/policy/defense/241447-pentagon-raises-protection-level-at-military-bases

http://savannahnow.com/news/2016-12-09/army-georgia-power-build-250-acre-solar-energy-farm-fort-stewart

http://www.autonews.com/article/20160822/OEM06/308229999/army-develops-autonomous-vehicles-for-use-on-bases-first-battlefields-

Media's Role in Energy Efficiency
by Yerika Reyes

Shifting buildings to become energy efficient is not sexy. Who wants to be lectured all about HVAC upgrades, incremental improvements to boiler systems, and LED lightbulbs? No one. However, what if it was in an Apprentice-style energy efficiency face-off? This is exactly what Better Buildings Challenge SWA does. They have two organizations try to find wasted energy in the other's operations, funded by the Department of Energy as a web series on YouTube. The Better Buildings program was started by the Obama Administration, and seeks to spread efficiency to the nation's commercial stock, including offices, schools, and city works.

In the show's first seasons, which happened last year, they had Whole Foods take on Hilton Worldwide. Their second season pitted the Air Force Academy against the Naval Academy. Now, their are turning to cities to face-off, with Boston taking on Atlanta. Teams from each city visit a variety of properties, including but not limited to airports, libraries, water treatment works, and streets. They find faults in corridor airflows, rhapsodize about rooftop solar panels, and needle each other about phantom light fixtures and the height of heating fans. Buildings in total account for at least one third of the nation's greenhouse gas emissions and are potential places for large amounts of savings if upgraded to be energy efficient.

Maria Vargas, head of the Department of Energy's Better Buildings program, which commissions the web series, would like the show to emphasize how organizations can learn from each other. "Hopefully by seeing real world situations and obstacles, SWAP provides a place for people to learn and feel empowered about making energy efficiencies," she told a reporter at Fast Company. She goes on to say, "[t]here's money investment involved here and people are naturally risk-averse. Seeing other cities do it, and seeing the investments work out, is good information to have." The show has been successful with about 2 million views on YouTube between season one and two.

Stephanie Stuckey-Benfield, Atlanta's chief resilience officer, appeared on SWAP and in an interview in the show expressed that she learned how Boston does procurement of energy-efficiency technology. She also expressed envy because Georgia doesn't have Massachusetts's regulatory framework, including energy-efficiency targets and a Renewable Energy Portfolio Standard that requires utilities to increase their use of renewables over time. Without these measure, funding projects in Atlanta is difficult. Additionally, Boston's chief of environment, energy, and open space, Austin Blackmon, explains the benefits of having an outsider inspect the cities work. During the show Atlanta folks identified unnecessary lighting in Boston's library buildings where natural daylight would suffice, for instance.

SWAP is not only granting access to information to the participants viewing the show and providing an invaluable experience that can boost energy efficiency, but also showcasing to viewers who might be skeptical the power of saving energy. New forms of media are making it increasingly accessible to understand energy efficiency. Tune in!

https://betterbuildingsinitiative.energy.gov/swap
https://www.eenews.net/stories/1060053325

"Free Energy Life" Offered in Sustainable, Luxury Tower

by Mary-Catherine Riley

Would you like to live in the lap of luxury in a flat northwest of London that provides excellent amenities and still feel like you are living sustainably? Excellent, so would I! For a starting price of £200,000, you too can live in the Beacon, "the world's most sustainable luxury tower."

The vision of the Beacon is to reinvent the modern lifestyle while incorporating the most technologically advanced sustainability practices. The Beacon integrates a multitude of renewable energy sources to offer a "free energy life". Beacon incorporates underfloor heating, heat recovery ventilation and geothermal heating to regulate the temperature of the building as well as smart thermostats to use that energy efficiently. The windows are three glazed and provide the best noise and heat insulation. The Beacon lowers the energy bill by using the power created by their 0.8 MW solar array. Furthermore, they have adapted motion sensitive LED light to provide the most efficient lighting.

The Beacon utilizes a 100% rainwater harvesting system to irrigate the plants located in the arboretum and on the balcony. This arboretum is used to freshen the air provided in the building using a natural ventilation system. Finally, the Skyline Parking structure provided by the corporation is a robotic parking system that returns your car in 45 seconds and eliminates the need for lighting and idling. The building has electric cars and electric bikes to use for those without cars.

The builders of the Beacon, Lumiere Developments, say that their model will act as a blueprint for "zero-carbon residential development" in the future. However, they point to the trends as three quarters of UK construction companies already use low-carbon strategies. As this technology continues to expand, it will become cheaper, and finally, the Beacon will not be an anomaly but a standard on how to incorporate luxury and sustainability.

(http://www.edie.net/news/6/Hemel-Hempstead-set-to-welcome--world-s-most-sustainable-residential-tower-/).

(https://www.thebeacondevelopment.co.uk/sustainability#sustainability2).

(http://www.edie.net/news/6/Hemel-Hempstead-set-to-welcome--world-s-most-sustainable-residential-tower-/). The Edie Newroom. "Hemel Hempstead Set to Welcome 'world's Most Sustainable Residential Tower'." Edie.net. Faversham House Ltd 2016, 21 Oct. 2016. Web. 20 Feb. 2017.

"The Beacon: Sustainability." Environmentally Sustainable Buildings | The Beacon Development. Lumiere Developments, n.d. Web. 20 Feb. 2017.

Vertical Agriculture: Inner City Farms

by Mary-Catherine Riley

Most conservative estimations say that there will be at least three billion more people by 2050. In reality, this means that a farm the size of Brazil will need to be added to the current quantity of farms in order to grow enough food for this increased population size. However, over 80% of the world's farmable land is currently in use. Furthermore, by 2050, approximately 80% of the Earth's population will reside in urban centers, increasing the distance between people and the production of their food. A potential solution to this is vertical farming; a method of growing crops, usually without soil or natural light, in beds stacked vertically inside a controlled-environment building. In 2016, Aerofarm built the world's largest vertical farm. It is in a converted steel mill and is a seventy thousand square foot floor space that allows for thirty-six foot tall columns of growing agriculture in eighty-foot-long rows. Dr. Despommier coined the phrase and frequently discusses the science and benefits of vertical farming stating, "If AeroFarm's vertical farm grows a thousand tons of greens a year, about fifty tons of carbon will be taken from the air".

Further benefits of vertical farming include drastically reducing runoff and the use of freshwater due to the precision and control allowed for in vertical farming. Runoff water is one of the main pollutants in the ocean and outdoor farming consumes 70% of the planet's freshwater.

The technology is partly derived from systems designed to grow food on the moon.

The space where the food is grown is very futuristic. From the outside, the mini farm looks like a metal trailer. Inside, there are rows of stacked plants basking in pink and purple lighting. The L.E.D.s have had their heat-producing part of the spectrum removed, which is the most energy intensive part of the spectrum, saving energy and increasing plant growth. Algorithms adjust the light to different intensities depending on the stage of plant growth.

This is an innovative method of growing food in urban areas that is beneficial to the environment and also brings fresh food to people in cities. However, it is an expensive proposition, at almost $4 for 4.5 ounces of salad. It will be interesting to see how urbanization, environmental conservation, and class intertwines in this complex problem.

(http://www.newyorker.com/magazine/2017/01/09/the-vertical-farm).
(http://www.newyorker.com/magazine/2017/01/09/the-vertical-farm).
(http://www.newyorker.com/magazine/2017/01/09/the-vertical-farm).
(http://www.verticalfarm.com).
Frazier, Ian. "The Vertical Farm." The New Yorker. The New Yorker, 09 Jan. 2017. Web. 14 Apr. 2017.

The "Greenwashing" of the United States Military

by Sara R. Roschdi

The U.S Military is the world's number one consumer of oil and producer of hazardous waste. The Union of Concerned Scientist report that the U.S. military consumes over 100 million barrels of oil to power naval ships, air force aircrafts, army tanks, vehicles and ground operations. Oil accounts for more than three-fourths of the departments of defense total energy consumption. Since the Cold War, the U.S. Military has reduced its consumption of oil yet the War on Terror has brought the usage to surge. The Department of Defense has boasted about its reduction in energy consumption, but this is largely due to the increased privatization of the military that has left energy consumption to the responsibility of private sub-contractors. The Department of Defense spends over $20 billion dollars on its total energy consumption. The United States Air Force is the largest consumer of energy consuming 2.5 billion gallons of fuel at an average cost of over $9 billion dollars per year. The Trump Administration plans to increase the military budget by 10% or $54 billion dollars to a total of $603 billion dollars. This is in comparison to the Department of Education total budget of $68 billion dollars and of that; we spend more on the military's energy than on Title I funding for disadvantaged students and for the Individuals with Disabilities Education Act for students in special education programs. The U.S. military with collaboration of conservationist are engaging in the ""greenwashing"" of the U.S. military seeking to improve the military's image by increasing its use of eco-friendly fuels and energy sources. This does not take into considerations the environmental impact of the devastation of militarism, bombs and army's bases to the land and people we are occupying.

https://consortiumnews.com/2016/09/09/greenwashing-wars-and-the-us-military/) .
https://www.electricchoice.com/blog/united-states-military-energy/) .
(http://oilprice.com/Energy/Energy-General/A-Look-At-US-Military-Energy-
    Consumption.html).
http://www.ucsusa.org/clean_vehicles/smart-transportation-solutions/us-military-oil-
    use.html#.WPRVzVPyt-U).
http://www.resilience.org/stories/2007-05-21/us-military-energy-consumption-facts-and-
    figures/
http://www.ucsusa.org/clean_vehicles/smart-transportation-solutions/us-military-oil-
    use.html#.WPRVzVPyt-U
http://www.resilience.org/stories/2006-02-26/us-military-oil-consumption/
http://www.guamagentorange.info/yahoo_site_admin/assets/docs/Environmental_Distruct
    ion_By_Military.238134937.pdf
http://blogs.edweek.org/edweek/campaign-k-
    12/2017/02/trump_education_department_cuts_proposed_budget.htm
https://consortiumnews.com/2016/09/09/greenwashing-wars-and-the-us-military/
http://oilprice.com/Energy/Energy-General/A-Look-At-US-Military-Energy-
    Consumption.html

Reflecting Heat into Space

by Justin Wenig

SkyCool Systems has developed a novel way to cool buildings by radiating heat into space. In March of 2016, the company demonstrated a prototype cooling unit which reduced the temperature of water to almost 5 degrees Celsius below air temperature. Most strikingly, the cooling unit used no water and little electricity. SkyCool Systems hopes its technology will be used in the near future as a supplement to commercial air conditioners.

The technological breakthrough is a novel coating material that promotes radiative cooling. Radiative cooling to the sky happens when visible radiation an object absorbs from the sun is reradiated as heat directs back into space, cooling the object. While scientists have known about radiative cooling for centuries, they have only been able to develop a commercial solution that works during the night time. SkyCool System's cooling panel represents a significant leap forward in technology, causing a sizeable amount of radiative cooling during the daytime as well as the nighttime.

The potential impact of a low-cost, energy and water-efficient cooling system cannot be understated. According to the Advanced Research Projects Agency- Energy (ARPA-E), the electricity required to cool buildings accounted for 14% of electricity use in the United States in 2015. Further, a 2016 study conducted by the Berkeley National Laboratory found that the world's supply of cooling units is projected to increase from 900 million in 2015 to 1.6 billion in 2030.

Clearly then, SkyCool System's technology has the potential to significantly cut electricity use for the rest of the 21st century. Founded in 2016, the company already has five million dollars in funding and a $550k grant from ARPA-E. If all works out as planned, the air conditioner as we know may change forever, and much for the better.

https://arpa-e.energy.gov/?q=slick-sheet-project/radiative-coolers-rooftops-and-cars
http://skycoolsystems.com/#home
https://tomkat.stanford.edu/innovation-transfer/skycool

Personal Air Conditioning from a Robot
by Justin Wenig

In February of 2017, researchers at the University of Maryland demonstrated a prototype personal air conditioner called RoCo. RoCo is a small, robotic air pump on wheels that tracks user movement and sends air at the desired temperature toward the user. When the user moves, RoCo follows along close behind and pumps air toward them. By avoiding the need for a central system, RoCo uses less energy than traditional air conditioners. Researchers are hopeful that the technology will be useful in low-density environments such as hospitals and offices in the future.

RoCo has to clear a few more technological hurdles before reaching the market. While the prototype was able to follow the user and pump air to the desired location, it could not detect user body temperature. To develop this technology, the researcher's plan is to build a small wearable device or smartphone app. Another difficulty is that the prototype was only able to operate for two hours before requiring a charge. However, according to Darryl Fears of the Washington Post, the researchers are confident that slight changes to the prototype will extend RoCo's battery life.

While RoCo is far from a safe bet, energy experts are optimistic the technology will come to fruition. In 2015, The United States Energy Department granted the University of Maryland 2.5 million dollars to develop RoCo. RoCo is also competing in a competition called Delivering Efficient Local Thermal Amenities (DELTA) that finances energy innovations deemed too risky for private sector investment.

If the researcher's vision for RoCo comes true, the technology could have a huge impact. According to energy.gov, almost 14 percent of U.S. energy output is used for air conditioning. The market for air conditioning is also exploding. A 2016 study from the University of Berkeley projects that the world is to install 700 million air conditioners by 2030. Hopefully RoCo can help satisfy some of that demand and save a lot of energy.

Smart Grid

Revamping of Energy Grids by the University of Toledo

by Sagarika Gami

The University of Toledo, in partnership with the U.S. Department of Energy, is making leaps in the realm of energy grids, shifting to transaction-based energy management. With climate change finally accruing noise in the media, more and more university towns are seeking to bolster the efficiency and reliability of their energy grids. The current grid system is very simple—the power companies match their production to the demand of homes and businesses and deliver based on these numbers. However, to keep up with demands for change in sources of power and its output, the grid system will likely become more complex.

This project works within the university's smart-building technology and one megawatt of solar generation capability. A new software, Volttron, will tap into the grid to collect information based on energy prices, generation, and demand. This information will be used to control how the buildings on campus can draw power and from where they will draw power – either UT's solar array or alternatively, from a battery that will collect roll-over power from the solar array, putting lower strain on the grid itself. The hope is that the grid will be able to automatically adjust energy loads based on the information gathered, developing mechanisms for minimal human direction. The transactive nature of this kind of energy management stems from how the quantity of energy usage in buildings on these campuses could result in more efficiency and reduced costs. This system will work to provide more flexibility for the power grid itself.

These changes will make the grid lower cost and more flexible in how it disseminates power throughout the campus. As one of the primary real-world tests, UT is working with the NASA Glen Research Center on the project with $1 million of funding from the Pacific Northwest National Laboratory. This project is just beginning and is a pioneer for this grid system. While UT and its collaborators are optimistic about the progress, we're keeping an eye out for bumps in the road.

GovTech.com (http://www.govtech.com/fs/infrastructure/Research-Project-Aims-to-Upgrade-Electric-Grid.html)

ToledoBlade.com (https://www.toledoblade.com/Energy/2017/01/22/Study-eyes-how-to-get-buildings-grid-to-talk.html)
Phys.org (https://phys.org/news/2015-11-bed-advances-washington-state-hotbed.html

Energy Monitoring Revamped

by Sagarika Gami

Mark Chung, an electrical engineer trained by Stanford University, began his venture butting heads with climate change seven years ago. It is widely thought that in order to combat the worst impacts of climate change, global carbon emissions must be cut by 40 to 70% by mid-century. Chung's company, Verdigris, seeks to aid the process by providing a solution to inefficient energy monitoring and usage. Verdigris came about as a response to "smart meters," which track where energy is being used in houses, buildings, hospitals, etc. The "smart meters," unlike Verdigris' software, are unable to create electrical maps on a large scale to monitor appliances, machinery, lights, and more, and are thus unable to pinpoint the exact sources of energy usage.

Verdigris, now backed by NASA, is able to isolate which sources use more and less energy by mapping every electron that a building uses. Chung created a clamp mechanism that attaches onto electrical wires to measure the flow of current around 8,000 times a second. Basically, these sensors are used to measure the electrical input of any machinery. These clamps collect information from a building's electrical panel and send the data to the cloud. From here, Verdigris integrates Artificial Intelligence to automatically phase out inefficient appliances by transforming factories and other buildings into living environments. Its technology makes the machinery smarter and more connected, taking the information gathered from the sensors mapping a building's electrical panel from which their AI is able to analyze the information, the "electrical fingerprints" of each machine, to optimize building controls, predict and prevent future breakdowns, and notify the company regarding energy usage. This technology helps to identify small areas of change, which in turn, adds up to big savings.

Investing in energy-efficient buildings is the cheapest and quickest method to cut energy use and greenhouse gas emissions, yet not many companies take this step. John Sternman (MIT's Sloan School of Management) attributes this to being a more psychological and political behavior as company attitudes are less inclined to fix something that is not "broken". Slowly but surely, Verdigris is expanding its reach and its mission, approaching its first 100 corporate clients this year.

Pri.org (https://www.pri.org/stories/2017-01-23/silicon-valley-engineer-quest-green-world-making-buildings-smarter)
kqed.org (https://ww2.kqed.org/science/2017/01/24/california-startup-saves-companies-millions-and-combats-climate-change/)
verdigris.co (http://blog.verdigris.co/verdigris-named-cb-insights-ai-100-list?linkId=33451523

The Future of Energy in Britain

by Cybele Kappos

As an energy crisis looms in the horizon of UK energy, the nation prepares for alternative energy sources. The BBC's Jane-Frances Kelly talks about the plan to build an interconnector, which is a large cable connecting Hampshire to the French coast. This project is completely privately funded and Aquind, the developer behind this, is currently working on a deal with the French. The purpose of this interconnector would be to supply energy in a time when coal-fired stations are slowly being phased out but the infrastructure for gas and nuclear energy is not yet equipped to take over. It is predicted that nuclear stations will take 10 years to be able to run. The interconnector is allegedly said to be ready to run in 2021 and the technology has already been tested out by Aquind. The cable is to play a significant role in the supply of energy across the nation. Moreover, it is said that it will ease the pressure on the UK grid and reduce the possibility of blackouts. The interconnector will be completely underground and undersea. It will not be visible to the public but it will provide up to 2GW of energy to almost four million homes.

The UK already has four interconnectors running from France, the Netherlands, Northern Ireland and the Republic of Ireland. However, there is still the need to meet the energy demands because the current energy system is strained, which is reflected by higher energy prices. Since France relies on nuclear energy, Lord Callanan- a non-executive director of Aquind- said that he hopes to see lower energy prices in the future of UK energy.

This is a smart, efficient, fast and economic response to the concern about meeting energy demands. It simultaneously seeks to revise the UK grid without compromising the energy supply by receiving the necessary energy from France. By continuing to move away from coal-fired stations, Aquind also hopes to reduce the carbon footprint per capita. This is a brilliant example of how private investors are contributing to the formation of the future of Britain's energy.

Kelly, Jane-Frances. UK to double French energy supplies with new cable. 13 June, 2016
http://www.bbc.com/news/business-36516585
http://aquind.co.uk/news/new-uk-france-interconnector/

Off-grid and Mini-grid Energy Production in Rural Tanzanian Communities Receives Grant from African Development Bank

by Genevieve Kules

Tanzania is challenged by poor energy access in rural parts of the country. Access to energy in these rural communities will require off-grid and mini-grid projects. The Tanzanian government has set a goal of establishing 1.3 million electrical connections by 2022, especially in rural areas. This would raise their connected population to about 35% from 20% in urban areas and 7% in rural ones.

Tanzania's energy goals come around the same time as the African Union (AU) and United Nations (UN) are setting goals for the continent and the world. The UN stated their goal of Sustainable Energy for All by 2030 and the AU began the Africa Renewable Energy Initiative in 2015. The African Development Bank (AfDB) has been supportive in achieving these goals. In January they approved a grant of $870,000 to support off-grid energy systems in Tanzania.

Some of the options already used in Tanzania for energy conservation and off-grid or mini-grid production are solar, multifunctional platforms, and improved mud stoves. Improved mud stoves have cut down on deforestation by decreasing the number of bundles of wood needed to burn a stove from 3 to 1, and increasing the efficiency of cooking on these stoves, although with the right technology you could probably charge your cellphone with energy produced by burning wood. Solar energy is the major off-grid / mini-grid energy generation method discussed so far. Multifunctional platforms are basically diesel engines that can power machinery, say for extracting palm oil, and produce energy to light about 250 light bulbs. Other off-grid energy production processes that have not yet been proposed for Tanzania include geothermal, wind, and micro hydro power. Many energy initiatives similar to the one by Tanzanian government are currently popping up across the continent of Africa and around the world.

Richards, James. "AfDB Support for Tanzania's Off-grid Energy Plans." Public Finance International. EY Building a Better Working World, 6 Jan. 2017. Web. 31 Jan. 2017. <http://www.publicfinanceinternational.org/news/2017/01/afdb-support-tanzanias-grid-energy-plans>.

Obiang, Olivia Ndong. "AfDB Scaling up Support for Africa's Development at AU Summit."African Development Bank Group. N.p., 30 Jan. 2017. Web. 31 Jan. 2017. <https://www.afdb.org/en/news-and-events/article/afdb-scaling-up-support-for-africas-development-at-au-summit-16683/>.

Cyber Security for Interconnected Energy Systems

by Genevieve Kules

Cybersecurity is increasingly important in today's internet of things: from protecting your passwords and bank accounts to protecting renewable energy grid systems. There are a number of ways for hackers and malware to be planted into an energy grid, and there are becoming more ways now that these systems have many different entry points for collecting energy.

Over the past fifteen years there have been numerous attacks, some examples of which are hacking, human error, malware, and viruses. According to the World Energy Council, some of these attacks have managed to shut down power in a city or town. Others have infected tens of thousands of computers, while some were done simply to show the weakness of a company or to gain access to private designs. According to the authors, cyber attacks are more likely and possible in developed countries with more energy infrastructure, because there are more input points and have more potential information stored within them. One major concern is the potential to cross from the cyber world into the physical one, such as widespread power outages. The authors also mention that centralized systems are at higher risk because of the possible "domino effect" from the central location.

The authors describe some preventative measures such as general education to those who are operating the system on any level. But because the technology is very new, there is not much public knowledge of the dangers of interconnected energy systems. Insurance is another financially protective measure that is becoming common in the cyber energy sector, though it has a ways to go because of the lack of historical knowledge about insuring this type of system.

Steel, William. "Heeding the Call for Cybersecurity in the Renewable Energy Sector." Renewable Energy World. N.p., 10 Apr. 2017. Web. 11 Apr. 2017. <http://www.renewableenergyworld.com/articles/2017/04/heeding-the-call-for-cybersecurity-in-the-renewable-energy-sector.html>.

"The Industrial Control Systems Cyber Emergency Response Team (ICS-CERT)." ICS-CERT. Department of Homeland Security, n.d. Web. 11 Apr. 2017. <https://ics-cert.us-cert.gov/>.

"WORLD ENERGY PERSPECTIVE: THE ROAD TO RESILIENCE - MANAGING CYBER RISKS." World Energy Perspectives: The Road to Resilience - Managing Cyber Risk | Swiss Re - Leading Global Reinsurer. Swiss Re, 27 Sept. 2016. Web. 11 Apr. 2017. <http://www.swissre.com/library/partner-publication/world_energy_perspective_the_road_to_resilience_financing_resilient_energy_infrastructure.html>.

Micro-grid in Brooklyn Encourages Community Regulated Energy Sharing

by Nina Lee

TransActive Grid, LO3 Energy's newest project, is connecting neighborhoods in Brooklyn by creating solar energy community micro-electrical grids. With solar panels installed atop residential, business, and industrial buildings across the span of ten blocks, the micro-grid collects solar energy and allows members to share it energy with each other. What usually happens when solar panel owners produce more energy than they need is that the surplus energy gets sent back to a central energy company, who then proceeds to sell that energy back to someone else on the grid. What TransActive Grid aims to do is cut out the centralized energy company and create a market controlled solely by consumers. That way, the funds used to buy energy from the community goes straight back into the community.

Using this decentralized business model not only saves money, but also minimizes the amount of energy lost when it travels back and forth from power plants. According to the U.S. Energy Information Association, transmitting energy long distances around the United States results in a 5% loss of energy. In the future, this setup could

267

also allow for excess energy to be donated to low-income areas. Another non-monetary benefit to having multiple decentralized power grids is that in the case of a natural disaster, grid malfunctions would not wipe out entire large communities at a time.

Unfortunately, this transaction cannot yet be completed with real money, since it is illegal for individuals to buy and sell electricity from each other without going through a utility company. Due to this law, all transactions thus far have used blockchain, a digital trading currency similar to bitcoin. Blockchain technology allows for transparent, traceable transactions between users and allows individuals to be fully in control of the finances surrounding their own solar energy production.

http://www.reuters.com/article/us-energy-usa-blockchain-idUSKBN171003
https://www.newscientist.com/article/2079334-blockchain-based-microgrid-gives-power-
    to-consumers-in-new-york/

New York Seeks to Revamp PowerGrid
by Kieran McVeigh

As they say change is the only constant. This phrase seems particularly true when it comes to technology. The state of New York is trying to keep up with the change in energy technology, by revamping its current power grid. New York plans to do this through the Reforming the Energy Vision (REV) model, which aims to create a new power system designed to deal with distributed technologies that are smart, and capable of interacting with the grid. At the center of this vision is trying to reward utilities based on the "value" they create for customers rather then necessarily the energy supplied.

A key component of this vision is to create economic incentives for utilities companies to maximize efficiency, and reduce their pass-through costs. This model may work very well in New York since the utilities do not own energy generation, but rather the infrastructure to transport the energy. What this looks like in practice is creating a more flexible grid, able to both store excess energy that may be generated by renewable energy sources, and when not much energy is being generated, supply places that normally generate energy with energy from else where.

New York plans to do this with some free market principles, by making distributed assets worth different amounts of money depending where they are. Basically ensuring that rather than giving people or companies incentives across the board to invest in solar panels or batteries, ensuring that, where-ever the need for solar panels or batteries is highest the incentives are highest. Once this system is in place, a system of granular price will be used to incentivize growth in areas where it is needed most. In this way the REV model emphasizes efficiency and alternative solutions rather than building power plants that are rarely run and costly to build.

The REV admittedly is a big and (complicated goal) but it also may lead the way for the reform of current energy systems across the US, and make a significant impact in the use of renewable energy sources.

Kelly-Detwiler, P. (2017, April 14). New York Intends To Revolutionize The Electric Power Grid As We Know It. Retrieved April 18, 2017, from https://www.forbes.com/sites/peterdetwiler/2017/04/14/new-york-intends-to-revolutionize-the-electric-power-grid-as-we-know-it/3/#32d4b7422863

Roselund, C. (2017, March 23). Christian Roselund. Retrieved April 18, 2017, from https://pv-magazine-usa.com/2017/03/23/we-are-at-war/

Vehicles

London Confirms New Charge on Polluting Vehicles

by Emily Audet

In February 2017, the mayor of London, Sadiq Khan, confirmed the establishment of a £10 fee, called a "toxicity charge" or "T-charge" for short, for the most polluting cars in the city. The charge applies to any vehicle falling below Euro 4 standards, which comprises mostly diesel and petrol vehicles produced before 2006. The charge will probably affect about 10,000 cars in London and the implementation will begin on October 23, 2017.

This T-charge is one of many efforts from Mayor Khan to reduce emissions in London. In November 2016, Khan claimed London would halt adding entirely diesel double-decker buses to its inventory by 2018 and, instead, the city would add zero-emission and hybrid buses. Khan describes the T-charge as a step towards the development of an Ultra-Low Emission Zone in London, which could be realized by 2019. Khan has also doubled the budget for air quality measures to £875 million over the coming five years. Khan cited strong public support and public health issues in his rationale for the T-charge. Over 9,000 Londoners suffer premature mortality annually related to air pollution.

Journalist Ed Wiseman argues that this charge will fall unfairly on London's low-income residents, who own older vehicles since they cannot afford to upgrade to newer, greener models. He argues that a £10 charge will do more to financially burden these households than it will to reduce emissions. London already implements a Congestion Charge of £11.50 Monday through Friday from 7 AM to 6 PM, and observers have pointed out that vehicles below Euro 4 standards on affected roads during these times will be charged a fee totaling £21.50. Londoners with minimum wages would have to work three hours to pay this combined congestion and toxicity charge. London is considering establishing a program where people can scrap their old diesel cars for a £3,000 stipend to apply to the purchase of a greener car. Wiseman, however, argues that this stipend would not be enough for very low-income households to afford a newer, more sustainable car.

[http://www.telegraph.co.uk/cars/comment/londons-t-charge-unfair-tax-citys-poorest-motorists/].

[http://www.fleetnews.co.uk/news/fleet-industry-news/2017/02/21/mayor-of-london-to-introduce-10-emissions-surcharge-in-october].

[http://www.cnbc.com/2017/02/17/london-introduces-12-daily-toxicity-charge-for-most-polluting-cars.html].

"Mayor of London to Introduce £10 Emissions Surcharge in October." Fleet News. N.p., 21 Feb. 2017. Web. 21 Feb. 2017. http://www.fleetnews.co.uk/news/fleet-industry-news/2017/02/21/mayor-of-london-to-introduce-10-emissions-surcharge-in-october

Frangoul, Anmar. "London Introduces $12 Daily 'Toxicity Charge' for Most Polluting Cars." CNBC. N.p., 17 Feb. 2017–17 Feb. 500. Web. 21 Feb. 2017. http://www.cnbc.com/2017/02/17/london-introduces-12-daily-toxicity-charge-for-most-polluting-cars.html

Wiseman, Ed. "London's T-Charge Is an Unfair Tax on the City's Poorest Motorists." The Telegraph 20 Feb. 2017. Web. 21 Feb. 2017. http://www.telegraph.co.uk/cars/comment/londons-t-charge-unfair-tax-citys-poorest-motorists/

Energy Innovation Predicts Large Environmental and Economic Costs if Trump Weakens Fuel Efficiency Standards

by Emily Audet

The Environmental Protection Agency (EPA) will begin a new review of the fuel efficiency regulations established under the Obama administration. These guidelines, formed in 2011, require automobile manufacturers to decrease carbon dioxide emissions in cars and small trucks so that fuel efficiency reaches an average of 54.5 miles per gallon by 2025. After Trump's election, the EPA quickly began its promised midterm review of the regulations on November 30, 2016. On January 12, the EPA closed its review without altering the standards established in 2011. The Alliance of Automobile Manufacturers argues that the EPA hurried through this midterm review and that a second review process is necessary, and in February 2017, the alliance asked the EPA to open up the review once more.

Journalists anticipated that on Wednesday, March 22, 2017—when Trump visited automobile firms in Michigan—Trump would disclose a plan to use this review process to weaken the 2011 regulations. While a review of these standards is definite, actually changing these guidelines will likely take years and several legal battles, as environmental interest groups and states try to halt changes through the courts. Revising these regulations gets further complicated by the many stakeholders involved, as the EPA, the Department of Transportation, and the state of California are all expected to be involved in changes.

Energy Innovation, a research and policy analysis firm, used the Energy Policy Simulator to predict the effects weakened fuel efficiency standards will have on the environment and the U.S. economy. The firm compared predictions into 2050 for Obama's fuel efficiency standards with predictions for a situation where fuel efficiency standards remain at the same level after 2021. Their analysis found that halting fuel efficiency at 2021 levels would increase emissions of carbon dioxide equivalent by 2,800 million metric tons more than the Obama administration's standards by 2050. Their model also showed a cost of about $400 billion to the U.S. economy under a lower fuel efficiency standards model. On average, keeping fuel efficiency standards at 2021 levels would cost each American household an average of $331 per year in increased gasoline consumption. The firm claims that the estimated $200 billion in costs to automobile manufacturers from higher fuel efficiency standards would be more than outweighed by overall savings in the economy from decreased gasoline consumption. An NPR piece by Bill Chappell and Richard Gonzales also noted that the production of electric vehicles relies a lot on government requirements, and in the absence of strong fuel efficiency standards, production of electric vehicles could decrease drastically.

[https://www.forbes.com/sites/energyinnovation/2017/03/15/rolling-back-fuel-efficiency-standards-would-cost-americans-800-billion-add-six-billion-tons-co2/ - 610b8e05e600].

[http://www.npr.org/sections/thetwo-way/2017/03/15/519037545/epa-reopens-u-s-rules-setting-vehicle-efficiency-standards-for-2025].

Chappell, Bill, and Richard Gonzales. "EPA Reopens U.S. Rules Setting Vehicle Efficiency Standards For 2025." NPR. N.p., March 15. Web. 21 Mar. 2017. http://www.npr.org/sections/thetwo-way/2017/03/15/519037545/epa-reopens-u-s-rules-setting-vehicle-efficiency-standards-for-2025

Orvis, Robbie. "Trump's Fuel Standards Rollback Will Cost Americans $370 Billion, Add Nearly Three Billion Tons CO2." Forbes. N.p., 15 Mar. 2017. Web. 21 Mar. 2017. https://www.forbes.com/sites/energyinnovation/2017/03/15/rolling-back-fuel-efficiency-standards-would-cost-americans-800-billion-add-six-billion-tons-co2/ - 610b8e05e600

Moving Towards Less Car Ownership?

by Alejandra Chávez

An interesting article by Alissa Walker explains how car ownership is likely to decrease in the United States, even with a "fossil fuel-friendly" administration in the White House. I initially found this hard to believe, especially after reading that the majority of Super Bowl ads in 2017 were purchased by car companies—for the sixth year in a row. The article also states that 2016 marked the year with the largest number of new cars purchased, vehicle-miles driven, and the most amount of gasoline consumed in a year. The Rocky Mountain Institute (RMI) even spotted trends that could demonstrate the United States heading in "the wrong direction," in terms lowering its fossil fuel dependency.

Despite all this information, the focus switches to a roadmap released by RMI that outlines some of the reasons why car ownership in the United States could decrease by 2020. One of the main factors is that there simply will not be as high a demand—increasing gas prices, public transportation improvements, and transportation network companies (TNCs) are all factors. RMI estimates that consumers could save $1 trillion per year if they did not own a car, which is parked "over 90 percent of the time" anyway. While TNCs are used by some Americans, the article explains that these car-sharing services need to be made more financially accessible to a larger portion of the population. Perhaps that is part of the reason why cars are still purchased at their current rate. Regardless, Walker highlights that the number of electric vehicles sold in the United States and the number of Americans living car-free are both rising, slowly, but rising. Thus, whether it is to cut costs or pollution, steering away from car ownership could be in our future.

[http://blog.rmi .org/blog_2016_09_23_Why_Peak_Car_Ownership_Isnt_So_Farfetched]
Walker, Alissa. "Car ownership may decrease in the U.S.—here's why." Curbed. 2017.
Curbed (http://www.curbed.com/2017/2/6/13428414/car-buying-electric-vehicles-uber-lyft)

Green Airlines the Future of Flight

by *Dominique Curtis*

Green Airlines are spreading from United Airlines in Los Angeles, JetBlue in New York, to Alaska Airlines in Seattle they are taking steps to reduce its carbon emissions, including increasing the use of sustainable alternative fuels, improving aircraft efficiency and making the infrastructure more efficient. Jim Lane (2016) announces that after six years of negotiations the governments meeting at the International Civil Aviation Organization (ICAO) have agreed on the design of a global market based measure of international aviation.

United Airlines is using sustainable aviation biofuel for many of its regular scheduled flights. If you're flying with United Airlines between Los Angeles and San Francisco your flight is using AltAir Paramount sustainable biofuel. United Airlines sign a contract agreeing to purchase 15 million gallons of sustainable biofuel that it will be using over a 3 year period. Bryan (2017) reports that United Airlines is hoping to reduce fuels with carbon lifecycles by 50 percent or more. United plans on doing this by doing a 30–40 percent blend with traditional jet fuel and AltAir renewable fuels.

United Airline isn't the only airline thinking of greener skies. JetBlue topped United Airlines by signing a contract agreeing to purchase 330 million gallons of renewable jet fuel with SG Preston that they plan to use over a 10 year span. The fuel setup will be 30 percent renewable fuel and 70 percent traditional fuel. In the article Lane states that based on a life-cycle analysis the production of the renewable jet fuel plants oil will create a targeted 50 percent or higher reduction in greenhouse gases emissions per gallon.

According to Renewable Energy World (2017) the US is putting funds toward the production of low carbon diesel and jet fuels. They plan to use low carbon fuels that come from industrial wastes that are underutilised to produce 3 million gallons of low carbon fuels per year. The hope is that all airlines will be using some mixture of renewable fuel by 2021 and reducing the emission of greenhouse gases in the process.

Bryan, Chelsea. "United Airlines Takes a Chance on Biojet Fuel." Runway Girl. N.p., 18 Jan. 2017. Web. 01 Feb. 2017.

Editors. "United Airlines Begins Regular Biofuel Use for Flights." Renewable Energy World. N.p., 14 Mar. 2016. Web. 01 Feb. 2017.

Lane, Jim. "Aviation Climate Agreement Reached at ICAO." Pardon Our Interruption. Biofuels Digest, 10 Oct. 2016. Web. 01 Feb. 2017.

Lane, Jim. "JetBlue Makes Record-setting 330-Million Gallon Renewable Jet Fuel Order."Renewable Energy World. Biofuels Digets, 22 Sept. 2016. Web. 01 Feb. 2017.

Putting Tesla to the Test

by Ethan Fukuto

The Aliso Canyon gas leak of 2015 in Los Angeles's San Fernando Valley caused not only an environmental crisis—fuel shortages affected the region's supply and source of energy. The crisis was a turning point for Southern California's energy industry, the start of an experiment in the use of batteries to meet energy demands. Tesla's contribution to the effort, 396 batteries at Mira Loma in the city of Ontario, went online on the 30th of January and is capable of providing power to around 15,000 homes for four hours. The batteries themselves are built at Tesla's Gigafactory in Nevada, and the company's process of vertical integration now means each component of the battery is built in-house. They are designed to store energy during the day and release at night during times of highest demand in the evening. California's increasing demand and funding for renewable energy projects allowed the Mira Loma project to come together in just a few months' time, with the threat of climate change and the impending closure of the last of California's nuclear plants pushing the industry towards alternative sources of renewable energy. Battery stations are an easy fit into Southern California's energy infrastructure and regulations, bypassing the environmental reviews, generators, and water and fuel supplies generally required for new power plants. Batteries are expensive, comparable to natural gas plants, however California's investment in renewable energy sources in the past, such as solar, has lowered costs for other markets. Investing in batteries follows California's future-facing model in energy, heading, ideally, towards a decreased dependence on natural gas. Though batteries are not entirely without their own set of problems: Samsung's recall of their Galaxy Note 7 phone brought to light the potential hazard for battery fires or explosions if not properly assembled or maintained. A battery project in Oahu, Hawaii, for instance, caught fire three times in 2012, and any incidents in California's battery stations may lead to decreased funding and support for such projects. Regardless, Tesla's investment in batteries and California's support point towards a more sustainable future.

Cardwell, Diane. "Tesla Gives the California Power Grid a Battery Boost". New York Times (30 January 2017). https://www.nytimes.com/2017/01/30/business/energy-environment/battery-storage-tesla-california.html

Cardwell, Diane & Krauss, Clifford. "A Big Test for Big Batteries". New York Times (14 January 2017). https://www.nytimes.com/2017/01/14/business/energy-environment/california-big-batteries-as-power-plants.html

Emissions from Volkswagen Scandal Will Have Fatal Effects in Europe

by Ethan Fukuto

In September of 2015, Volkswagen admitted to cheating emissions regulations after an EPA investigation into VW vehicles noted higher-than-reported levels of nitrogen oxide (NO_x) emissions. Software in VW's diesel cars, known as a "defeat device", activated the full emission-control system only when vehicles were being tested—otherwise they ran without it. Use of the defeat device allowed the company to meet emissions standards when testing, but not so in real-world conditions. The EPA's findings only covered the United States, but surveyed over 482,000 cars which include VW-manufactured vehicles such as Audis and Porsches. Shortly after the EPA report, VW America officials apologized and agreed to pay up to $14.7 billion to settle claims.

Issues stemming from the scandal are much more pressing in Europe. A March 2017 study by MIT notes that 11 million cars sold in European markets between 2008 and 2015 have defeat devices built in—with 2.6 million alone in Germany. Tests conducted by MIT concluded that the mean NO_x emissions from these cars, 0.85g km^{-1}, is more than four times the European limit of 0.18g km^{-1}. The research team determined excess NO_x levels in Germany by aggregating data on VW's sales, driving behavior and on-road measurements. The team estimates that the excess emissions from VW cars with defeat devices will cause 1200 premature deaths, which corresponds to 13,000 life-years lost and €1.9 billion in costs. Around 500 deaths are predicted within Germany's borders, with the rest in neighboring states such as Poland, the Czech Republic, and France. In comparison, a 2015 study estimated 59 premature deaths in the United States due to VW vehicle emissions. The research team attributes Europe's higher population density and more NO_x-sensitive environment, as well as higher VW sales in Germany and a higher rate of annual kilometers driven per vehicle, to the large difference in mortality rates between the regions. If the new fleet of VW vehicles reduce emissions to European standards by the end of 2017, the research team estimates that around 29,000 life-years and €4.1 billion in medical costs can be saved. While defeat devices have now been outlaws in the United States, they still do not technically violate European law. But as the number of deaths and economic costs suggest, tighter regulations are needed to ensure a healthier future.

Chossière, Guillaume P., Robert Malina, Akshay Ashok, Irene C. Dedoussi, Sebastian D. Eastham, Raymond L. Speth, and Steven RH Barrett. "Public health impacts of excess NOx emissions from Volkswagen diesel passenger vehicles in Germany." Environmental Research Letters 12, no. 3 (2017): 034014.

Hotten, Russell. "Volkswagen: The Scandal Explained". 10 December 2015. BBC News. http://www.bbc.com/news/business-34324772

Tabuchi, Hiroki & Ewing, Jack. "Volkswagen to Pay $14.7 Billion to Settle Diesel Claims in U.S." 27 June 2016. New York Times. https://www.nytimes.com/2016/06/28/business/volkswagen-settlement-diesel-scandal.html

Yin, Steph. "Volkswagen's Emissions Fraud May Affect Mortality Rate in Europe". 6 March 2017. New York Times.

https://www.nytimes.com/2017/03/06/science/volkswagen-emissions-scandal-air-pollution-deaths.html

Barcelona Announces Biggest Car Removal Program in Europe

by Genna Gores

In early March, Barcelona announced the most comprehensive car removal program in all of Europe. By 2019, all cars built before 1997 and trucks built before 1994 will not be permitted to drive in the city on weekdays. On top of this removal program, the city will also create a focus group to monitor the reduction of air pollution and how it affects public health. By creating these new regulations the city will take 106,000 cars off the road—7% of the whole fleet in Barcelona—and 22,000 vans. Along with a congestion charge during busy traffic times and a new fuel tax for public transit, the Barcelona program will be a more extensive overhaul of the car fleet than in Paris and London, who are rolling out similar programs.

This change in Barcelona's car regulations will have major effects on the pollution emitted by single-person vehicles. Since this program covers 40 municipalities, beyond the city center, it could have a huge effect on NO_2 emissions. This type of reduction is important because Barcelona pledged to reduce their emissions by 30% over the next 15 years. Car removal will also benefit public health. In a recent statistic, 3,500 people die a year in Barcelona due to pollution-related health problems, and the removal will hopefully mitigate that. While car removal is overall a good idea, it does disproportionately affect those who cannot afford new cars. To offset this, the Catalonian government is offering subsidized public transit passes for three years to those who give up their old cars. These subsidies will not only offer transportation for those who stop using their cars but also encourages the use of mass transit.

Beyond car removal, Barcelona will also transform many of their city blocks into no-car sections called Superilles. These will be sections of streets that will not allow through-traffic for cars and only be accessible for pedestrians, buses, and cyclists. With these changes Barcelona will free up almost 60% of their streets currently used for cars and give them back to citizens, and it will also reduce traffic by 21%. Barcelona regularly fails air quality targets and by removing old cars and streets for personal vehicles the hope is that they will meet their 15-year goal. Not only will these changes help reduce emissions, but the government hopes they will encourage a more mobile lifestyle and eliminate traffic accidents.

[https://www.citylab.com/commute/2017/03/barcelona-will-ban-older-cars/518805/]
https://www.theguardian.com/cities/2016/may/17/superblocks-rescue-barcelona-spain-plan-give-streets-back-residents
https://www.citylab.com/commute/2017/03/barcelona-will-ban-older-cars/518805/

Uber will Resume Self-Driving Car Program After Crash in Arizona

by Genna Gores

On March 24, 2017 one of Uber's self-driving cars got in an accident with another vehicle in Tempe, Arizona. This accident occurred due to the human-driven vehicle failing to yield to the autonomous car. The Uber car flipped with two safety engineers inside, but there were no paying passengers in the car or serious injuries. This caused Uber to suspend their self-driving program temporarily, but a few days later they resumed their testing. This is the first major crash involving an Uber autonomous car, but Uber's cars have also run red lights and had questionable run-ins with cyclists. While everyone involved in this crash ended up being fine, this accident shows that the transition period of human drivers to self-driving cars will not be an easy one. While autonomous cars could potentially be safer they do not account for human error that occurs around them.

With big companies like Uber, Google, and Facebook rolling out autonomous cars, the way that the world uses cars may change dramatically. Car-sharing programs, like Uber, could lower their prices because they are cutting out the cost of a driver. Ultimately this fare drop could encourage people to stop buying their own cars and just use car sharing programs—eliminating emissions from excessive personal car use, and particularly searching for parking spots. On top of lowering emissions, if the fleet of cars in general switches to autonomous vehicles it could potentially lead to fewer traffic accidents and safer streets. Uber, though, is facing many problems in their development of a self-driving cars. Beyond accidents, Uber is currently being sued by Google for stolen technology, and were told to stop testing it. Uber wants to be the leader in autonomous cars, and if Uber can bypass these bumps in the road they could change how people use cars for forever.

http://www.usatoday.com/story/tech/news/2017/03/27/ubers-self-driving-program-veering-off-track/99696940/

https://www.washingtonpost.com/news/innovations/wp/2017/03/27/uber-self-driving-cars-back-on-the-road-in-san-francisco-after-temporary-suspension/?utm_term=.1acdb27aeedd

Zunum Aero Could Change the Game in Air Travel

by Genna Gores

Zunum Aero, an aviation start-up backed by Boeing and JetBlue, is creating a series of small hybrid and electric planes that they plan to roll out in the early 2020s. According to Zunum Aero, 40% of pollution comes from regional flights, like San Francisco to Los Angeles, and with their new aircraft they could cut carbon emissions by utilizing regional airports instead of going through massive hubs, like JFK in New York city or LAX in Los Angeles. The goal is also to cut travel time with these aircrafts for example, they could fly from Sillicon Valley to Los Angeles in two and a half hours—shorter than traveling by car, which usually takes five to six hours. According to Business Wire, Zunum Aero will decrease door-to-door travel times on busy corridors by 40%, and lower operating costs cutting fares 40%–80%. The hybrid planes will also lower emissions by 80% and hopefully they will achieve zero emissions as battery technology improves. Lastly these planes will eliminate community noise by 75% by removing more commercial airlines and cars from Americans' travel plans.

Battery technology is advancing rapidly, and with groups like Solar Impulse that flew a solar powered aircraft around the world, Zunum Aero's goal seems even more plausible. Their aircrafts will carry 10–50 passengers, and will fill a gap in the market by using the underutilized regional airports. Currently, the air industry has most of their flights concentrated in 2% of airports, so Zunum Aero wants to change this. With small aircrafts providing flights between 700 to 1000 miles, they can eliminate the regional flights that congest major airports, and typically burn less fuel then long, international, flights. Zunum Aero want to make travel fast and affordable, while also considering the environment. Due to their quick progress, efficient team of engineers, and huge backings from Boeing and JetBlue, it is possible that within 10–15 years these types of planes will dominate short distance travel.

[http://www.businesswire.com/news/home/20170405005365/en/Zunum-Aero-
          Developing-Hybrid-Electric-Aircraft-Fast-Affordable]
http://www.businesswire.com/news/home/20170405005365/en/Zunum-Aero-
          Developing-Hybrid-Electric-Aircraft-Fast-Affordable
https://www.forbes.com/sites/grantmartin/2017/04/06/zunum-aero-could-be-the-tesla-
          of-aircraft/#701661ee5527

Big Goals for Small Islands: Hawaiian Legislation Promotes Renewable Energy in Transportation

by Siena Hacker

For such a small chain of islands, the state of Hawaii has the biggest renewable energy target in the United States. Hawaii introduced legislation that would fine utilities that are not completely powered by renewable energy by 2045. Now, as reported in a January 2017 New York Times piece by the Associated Press, the state is going a step farther by introducing legislation promoting a complete reliance on renewables for the transportation sector.
[http://www.nytimes.com/aponline/2017/01/19/us/ap-us-renewable-energy-transportation.html] With Hawaiians already owning an estimated one million cars – not to mention all of the cars for sale in dealerships – it would be imprudent for the state to mandate a shift to renewable fuels for the transportation sector. Hawaii is instead

attempting to encourage the transition by increasing the number of required charging stations. The reasoning holds that as electric cars become cheaper and the infrastructure supporting them increases, investing in an electric car will become the practical choice. Having spent the summer with family in Kauai, I was confronted with gas prices that were consistently a dollar higher than in California because of the cost to ship fuel. Those who currently own electric cars definitely see large savings, especially if they have their own solar panels and can charge their car during the day. Though there are only about 5,000 electric cars on the road in the state, Hawaii still ranked second in the nation in 2015 for the greatest number of electric cars relative to their population. There is abundant potential for renewable energy on the island to power transportation in the future, and the State has created a new position within their Department of Transportation to tackle this challenge. According to a December 2016 Biz Journals article by Duane Shimogawa reporting on the Hawaii Energy Resources Coordinator's Annual Report, Hawaii already gets 25%of its total energy from renewables. The Kauai Island Utility Cooperative alone saw an increase of 9.8% in its renewable energy, consistent with an upward trend in renewable energy production across the state. However, fossil fuels usage for commercial aviation remains a problem. Given the distance between Hawaii and the continental United States and the limitations of today's technology, airplanes cannot currently depend on batteries or other renewable energy sources. Advances in aviation technology will certainly be necessary for the state to its goal of complete dependence on renewables. However, a first step might be transitioning to renewables fuels for airplanes flying inter-island trips.

[http://www.bizjournals.com/pacific/news/2016/12/27/hawaii-achieves-25-clean-energy-in-2016-report.html]
Bizjournals.com http://www.bizjournals.com/pacific/news/2016/12/27/hawaii-achieves-25-clean-energy-in-2016-report.html
Nytimes.com http://www.nytimes.com/aponline/2017/01/19/us/ap-us-renewable-energy-transportation.html

Travel by Wind on Dutch Electric Trains

by Siena Hacker

Passengers on Dutch electric trains may not have noticed anything different about their rides during the first month of 2017, but they were actually being carried by wind--wind energy, that is. On January 1, 2017, the Netherlands' major national railway company, Nederlandse Spoorwegen (NS), announced that all of their electric trains now run on wind-generated energy. These trains also use regenerative braking, similar to hybrid cars, which produces energy while braking. This energy can power the train that generated it, or it can power other trains coming along the same line later. NS originally aimed to power all their electric trains with wind energy by January 2018 but were actually able to achieve their goal a year early. According to January 2017 Forbes article, the company serves about 600,000 people per day and requires about 1.2 billion kilowatt-hours (kWh) of electricity per year. According to a January 2017 article by The Guardian, one windmill generates enough energy within an hour to power a train for 120 miles. NS contracted Dutch renewable energy supplier Eneco two years ago for wind-produced electricity. By 2020, NS and Eneco aim to reduce the amount of energy utilized per passenger by 35% compared to 2005 levels. Eneco is also a member of a Shell-led consortium that recently won a bid to build a 700 megawatt wind farm off the southern Dutch coastline. A December 2016 Phys.org article reports that the wind farm, which carries an estimated cost of $319 million, is one of five farms to be operational by 2023. Each wind farm will generate enough energy to power one million homes. These wind farms will help the Netherlands reach its country-wide goal of deriving at least 14% of its energy from renewables by 2020. Hopefully these wind farms will also continue to power NS's electric trains for years to come!

Forbes: http://www.forbes.com/sites/lauriewinkless/2017/01/12/dutch-trains-are-now-powered-by-wind/#53450c964fe7

The Guardian: https://www.theguardian.com/world/2017/jan/10/dutch-trains-100-percent-wind-powered-ns

Phys.org: https://phys.org/news/2017-01-dutch-powered-energy.html

Technology and Policy Driving Down the Price of Electric Vehicles

by Siena Hacker

Though many people want to own electric vehicles (EVs), the price remains prohibitively high for most models. Tesla, the EV giant headed by Elon Musk, has been aiming to sell a model for $35,000. Although this price would make the car affordable to the average American, Tesla has been unsuccessful in reaching their target. One major reason is the high cost of producing battery cells and packs. According to a 2017 Futurism article, Tesla reports that their Gigafactory, which opened in Nevada in 2016, has reduced the cost of producing a battery by 35%. This reduction in cost is five % more than their initial estimate. EVs are already cheaper to own over their entire lifetime than traditional gasoline-using cars. However, the sticker shock of an EV prevents many people from buying them. Lowering the initial cost will allow more consumers to purchase these cars, giving them gas savings and reducing the amount of carbon dioxide emitted into the atmosphere.

Another way to reduce the price of EVs is through government incentives, such as rebates and subsidies. In our current political climate, many worry that federal incentives for buying EVs will cease to exist. However, many states are taking it upon themselves to encourage consumers to purchase EVs. For example, New York Governor Andrew Cuomo announced a new rebate program for up to $2,000 on March 3, 2017. According to a 2017 Electrek article, State assemblywoman Amy Paulin said that rebate would help make EVs "a mainstream option." Along with current federal incentives, New York EV buyers will receive about a $10,000 discount on their purchase. Even if federal incentives are discontinued, technological advancements and state policy initiatives will still encourage some consumers to purchase EVs.

Futurism: https://futurism.com/2-tesla-pulls-down-battery-cost-by-35/
Electrek: https://electrek.co/2017/03/03/new-york-state-ev-incentive/

Electric Bicycles Wheely Taking Off

by Siena Hacker

The technology for electric bicycles, or e-bikes as they are often called, has existed for at least a decade. However, e-bikes have exploded in popularity in recent years. E-bicycles allow riders to exert less energy while traveling at higher speeds, usually up to 25 miles per hour. However, e-bikes still give riders the option to get some exercise in on their route. Electric bicycles are usually pedal-assisted: bikers ride normally but receive a boost for each pedal. The majority of these bikes also have the option to select how much, if it all, the boost assists. There is a plethora of electric bicycle offerings from companies like Specialized, Volton, and Izip that cater to different audiences. For example, it is possible to choose from folding electric bikes, electric mountain bikes, or electric cruisers. E-bikes make longer bike routes more feasible and provide great environmental benefits due to decreased carbon dioxide emissions and reduced vehicular traffic. However, one of the biggest criticisms of e-bikes is that they are much heavier than traditional bikes and thus have a different, clumsier feel. A new Seattle-based bike company, Propella, is hoping to create a bike that is a better balance of e-bikes and traditional road bikes. The company recently launched an Indiegogo campaign to crowdfund "2.0," its second-generation e-bike. Its previous model was successful, but Propella is hoping to push more boundaries with 2.0. For starters, 2.0 will be 8% lighter than the first version due to a new aerodynamic frame, a lighter battery, and increased usage of alloy material. At just 34 pounds, it will still be about 14 pounds heavier than most traditional road bikes. However, it will be considerably lighter than most e-bikes, which usually weigh 50 pounds. This new model can reach a maximum speed of 20 miles per hour and can go a maximum of 40 miles before needing to be recharged. Propella has already raised 75% of its $60,000 crowdfunding goal. If they successfully raise the remaining $15,000, 2.0 will go on sale in September with an estimated retail price of $1,500 for the single-speed model and $1,650 for the 7-speed version.

Geeky Gadgets: http://www.geeky-gadgets.com/propella-2-0-lightweight-e-bike-05-04-2017/
Digital Trends: http://www.digitaltrends.com/outdoors/propella-2-e-bike-cycling-outdoors/

The Standards of Self-Driving Cars
by Cybele Kappos

Automakers are conflicted about how to make self-driving cars available for their customers in the future. Quain explores the advantages and disadvantages of having a standardized platform for the technology of self-driving cars. According to Nvidia, a technology company that is currently working with multiple automakers, the plan is to have self-driving cars on the road by 2020. Several car companies have expressed interest in this idea because a centralized approach would reduce the complexities of constructing the technology for the cars. It would allow huge cost savings by reducing the number of processors in the car and the amount of cabling among the processors. This would consequently lower the weight of the car and improve fuel economy.

Developing new software and safety systems could be easier under a common automotive platform. If there is no such platform, then the unique assembly of electronics, sensors and microphones in each car would have to involve individually tailored software for each model.

Although several car companies are collaborating with tech companies, automotive culture is known for its independence. Companies like Hyundai have developed their own platform that aims for affordability. To companies like this, a common platform would hinder innovations as well as potentially present the risk of a vulnerability in one car being a vulnerability in all cars.

Quain argues that "variety is not a virtue" when it comes to safety because these cars must interact predictably with human drivers and other robotic vehicles.

If there is a variety in how they behave, it could pose a risk. Equipment that could ensure communication from car to car, also known as the vehicle-to-vehicle or V2V communication, would allow them to share information about road conditions that sensors cannot detect.

Still, self-driving cars are new to the industry and every car company has to establish its own standards before discussing a common platform.

Quain, John R.. Self-Driving Cars Might Need Standards, but Whose. 23 February, 2017
https://www.nytimes.com/2017/02/23/automobiles/wheels/self-driving-cars-standards.html?rref=collection%2Fsectioncollection%2Fpersonaltech&action=click&contentCollection=personaltech&region=stream&module=stream_unit&version=latest&contentPlacement=2&pgtype=sectionfront
http://www.ubergizmo.com/2017/02/gm-toyota-lyft-standardized-self-driving-rules/

Hydrogen Cars: Open Roads but with out Enough Filling Stations
by Kieran McVeigh

This weekend as I zoomed along the ten freeway passing thousands of cars I found myself wondering what happened to

Hydrogen fuel cars that in the early 2000's were heralded as the next big thing. A late January article in Bloomberg technology gave me my answer, hydrogen fuel cells cars are alive and well. Many car manufacturers are preparing to or have already rolled out commercially available hydrogen fuel cars but these cars face major logistical hurdles because of the lack of available hydrogen fuel stations.

With the introduction of Toyota's Mirai, hydrogen fuel cell cars became commercially available in 2016, however Hydrogen comes at a price as a Mirai starts at about 60,000 dollars. Toyota currently only makes 3,000 Mirai a year so if demand and production ramp up this price will likely decrease. The major hurdle more then the relative expense of hydrogen fuel cell cars is the lack of network of filling stations. California leads the way with a total of 100 hydrogen fuel stations. Hydrogen fuel manufacturers insist that government subsidies are necessary for hydrogen fuel infrastructure to be completed, saying the costs of creating hydrogen fuels stations currently outweigh the benefits. As the all-too familiar problem surrounding global warming of how to get people take responsibility for our planets wellbeing when it will cut into their pocket books.

With the Trump administration officially having taken office in January of 2017, there seems to be little hope of establishing the infrastructure necessary to make hydrogen fuel cars a viable way to combat the emissions of greenhouse gases in the US. Although the assistance of the US government for hydrogen fuel cell cars seems unlikely, there are still some rays of hope. The governments of both the UK and Germany are investing in the creation of hydrogen filling stations to support the growth of hydrogen cars. Also eight privately owned car manufacturers and energy companies have banded together into an alliance to promote the increased use of hydrogen fuel cars. Will these new initiatives be enough to put hydrogen cars into gear? Only time will tell.

Lippert, John. "5 automakers form hydrogen alliance with energy, transport giants." Automotive News. N.p., 18 Jan. 2017. Web. 23 Jan. 2017.

Ma, Jie. "Toyota Chairman Says Fuel-Cell Cars Need More Time to Catch On." Bloomberg.com. Bloomberg, 19 Jan. 2017. Web. 23 Jan. 2017.

Wall, Matthew. "Hydrogen, hydrogen everywhere..." BBC News. BBC, 26 Mar. 2015. Web. 23 Jan. 2017.

Woodyard, Chris. "Review: Toyota Mirai dresses up, but hard to fill up." USA Today. USA Today, 11 Dec. 2015. Web. 23 Jan. 2017.

Scotland, London Propose New Low Emission Zones

by Kieran McVeigh

Scotland may become the latest country in Europe to institute low emission zones in their major urban centers. In January of 2017 the Scottish ministers proposed piloting a "low emissions zone" in the most polluted areas of Scottish cities to reduce pollution and help meet Scotland's greenhouse gas reduction goals. These goals are among the most aggressive in the world with an end goal of an 80% reduction in green house gases by 2050. The proposed low emission zone would prevent vehicles that create higher then average amounts of pollution like trucks or "lorries" as they're referred across the pond, from driving in the low emission parts of the city.

Low emissions zones have been instituted in cities across Europe, with low emission zones in London snagging the most the press. The current bans in London, ban all diesel cars that do not meet current environmental emission standards, and include a 10 euro charge for those that break the ban. In November of 2016 the high court in London ruled current low emission zones were not stringent enough and that more vehicles need to be banned to meet London clean air laws.

The court's ruling that London needs to strengthen its ban on diesel vehicles has unique political concerns surrounding it as the ban affects many families and small business owners, who, under the previous government, were encouraged to buy diesel cars and will now be subject to taxes for the cars they drive. Opponents of this ruling argue that the ban is ineffective, and that diesel cars, the main targets of the ban, produce as little as 5% of the pollution in the area. They suggest these resources would be better implemented modernizing the public transportation system to create less pollution.

Bans or taxes like those implemented in London and Scotland offer an interesting change in the economic incentives for those that own cars that pollute more. I believe economic incentives like these are a necessary part to combatting climate change as there is a large social loafing problem where there is no penalty for contributing more to greenhouse gas emissions. Although I believe it be a net gain, those who are affected by the change in policy of one government after just being incentivized to do the opposite by the previous government should be given assistance to comply to the new rules.

Reporters, Telegraph. "Scottish drivers face pollution charges in battle against climate change." The Telegraph. Telegraph Media Group, 20 Jan. 2017. Web. 30 Jan. 2017. <http://www.telegraph.co.uk/news/2017/01/20/scottish-drivers-face-pollution-charges-battle-against-climate/>.

Morgan, Tom. "Diesel drivers fear £10 city centre bill after High Court orders Government to clean up act on emissions." The Telegraph. Telegraph Media Group, 2 Nov. 2016. Web. 30 Jan. 2017.

Scottland http://www.telegraph.co.uk/news/2017/01/20/scottish-drivers-face-pollution-charges-battle-against-climate/

UK/England http://www.telegraph.co.uk/news/2016/11/02/diesel-drivers-fear-10-city-centre-bill-after-high-court-orders/

Two Companies Innovate Electric Buses in the United States

by Nadja Redmond

Transit vehicles are mostly powered by unrenewable power sources, such as gasoline, compressed natural gas (CNG), or diesel, with batteries only encompassing 1% of the market. Bus manufacturer

Proterra claims that its Electric transit buses are cheaper than the alternative diesel and CNG options. It's CEO, Ryan Popple, is making predictions that, in the next 10 years, electric transit buses powered by renewable energy will dominate the market. Specifically, he predicts that the majority of bus sales will be electric by 2025, and all new bus sales to transit agencies will be electric by 2030. King Country Metro Transit signed a deal for 73 buses with the company for use in and around the Seattle area. These buses can travel 23 miles between charges, with charges taking 10 minutes or less.

Microvast is another company that specializes in battery storage solutions; it has delivered more than 7,500 all-electric and hybrid-electric buses in 2016 alone, with 15,000 overall produced at the end of the year. Most of these buses are in use overseas in China and various European countries, and the company plans to expand operations to the US this year.

Microvast is a developer that manufactures long-life, fast-charging, lithium ion battery systems for more than 10 years, with their electric and hybrid bus business starting in 2010. The U.S. headquarters is based in Houston, and previously focusing efforts overseas, Microvast recognizes a monumental opportunity for growth and innovation in the US.

[http://metro.kingcounty.gov/am/innovations/examples/battery-buses.html]
[https://electrek.co/2017/02/13/electric-buses-proterra-ceo/].

The Crib that Simulates a Car Ride

by Yerika Reyes

New parents are always seeking ways to get more sleep and willing to try anything, especially if it will save them money. Ford noted in a press release that new parents can expect to lose the equivalent of 44 days of sleep in the first year of their child's life alone. A trick that parents have used to put their babies to sleep: long car rides.

The repetitive passing patterns, subtle vibrations, the hum of the engine help to soothe the baby to sleep. Ford has decided to design a crib that can simulate all those experiences called the Max Motor Dreams, without the need for a driver or any gas. "After many years of talking to mums and dads, we know that parents of newborns are often desperate for just one good night's sleep, said Max Motor Dreams designer Alejandro López Bravo, who helped create the design. "But while a quick drive in the family car can work wonders in getting baby off to sleep, the poor old parents still have to be awake and alert at the wheel. The Max Motor Dreams could make the everyday lives of a lot of people a little bit better."

The crib looks like a normal baby bed, but thanks to a companion smartphone app, it's capable of much more. The crib contains LED lights that glow similarly to street lights, and has speakers at the bottom that can make muffled engine sounds for ambient noise. It gently vibrates and rocks to mimic a ride in the backseat, and even comes with an app designed to track your car's different route so it can reproduce the movements from that drive for your baby. The Max Motor Dreams appears to be mostly conceptual for now, with only one crib produced so far, but if there's enough demand for the smart crib, the automaker says it will consider a full production run in the future. No pricing information is available as of now, but Ford España is running a sweepstakes to give away the crib to one lucky winner, if they partake in a test drive of the Ford Max.

The crib was designed by ad agency GTB, working with Ogilvy and Mather, to help promote the Ford C-MAX and S-MAX in Spain – two models quite likely to be purchased by folks with kids. But following the response Ford has received from interested parents, the Max Motor Dreams could make it to scale production. The hope for parents is not only could this come crib become true and that can get some more sleep, but hopefully save money on the road.

http://www.businessinsider.com/ford-baby-best-crib-simulates-driving-the-motor-dreams-2017-4

http://www.cnbc.com/2017/04/10/ford-crib-mimics-car-motions-helps-babies-sleep.html

Making Public Transportation a Zero-Emission Service

by Mary-Catherine Riley

In April of 2016, New Flyer Industries Canada ULC released the Xcelsior ® XHE60 heavy-duty articulated fuel cell transit bus during their inaugural road demonstration. The zero emission New Flyer XHE60 fuel cell bus is 60 feet long and carries over 120 passengers. The mechanics are combinatory. While the bus is operating, a fuel cell is used to provide a steady source of electricity while the batteries capture the breaking energy and provide propellant for acceleration. The combination of battery, fuel cell, and hydrogen storage gives these buses a driving range greater than 250 miles without recharging or refueling.

However, this is not even close to the first zero-emission bus New Flyer Industries has produced. In fact, this company is not only the leading manufacturer of heavy-duty transit buses and motor coaches in the United States and Canada, but it also produced 83% of the equivalent units (EU) of zero-emission heavy-duty transit deliveries in 2016. Moreover, New Flyer manufacturers three types of zero-emission buses including battery-electric, trolley-electric, and hydrogen fuel cell electric buses in order to optimize the served location's infrastructure, changing methods, and range capabilities.

"A zero emission transit bus (ZEB) is able to eliminate 1,690 tons of CO_2 over its 12-year lifespan which is equivalent to taking 27 cars off the road for each ZEB. Additionally, each ZEB also eliminates approximately 10 tons of nitrogen oxides and 350 pounds of particulate matter, improving air quality in the communities served." These initiatives of reducing overall vehicle emissions are essential in aiding cities to meet national air quality standards and lowering greenhouse gas emissions while still providing citizens mobility.

(https://www.newflyer.com/rss/802-zero-emission-new-flyer-bus-deliveries-increase-by-48-in-2016).
(https://www.newflyer.com/rss/699-new-flyer-announces-the-debut-of-the-first-hydrogen-fuel-cell-60-foot-bus-in-north-america).
Koffman, Jon. "New Flyer - New Flyer Announces the Debut of the First Hydrogen Fuel Cell 60-Foot Bus in North America." New Flyer. New Flyer Industries, Inc., 18 Apr. 2016. Web. 02 Feb. 2017.
Koffman, Jon. "New Flyer - Zero Emission New Flyer Bus Deliveries Increase by 48% in 2016." New Flyer. New Flyer Industries, Inc., 11 Jan. 2017. Web. 02 Feb. 2017.

Autonomous Underwater Vehicles Are Becoming Self-Sustaining as Well

by Mary-Catherine Riley

The average cost per day of a research vessel at sea is enormously high. Therefore, there are efforts to explore the deep blue sea in more economical manners. This is where autonomous underwater vehicles come into play. They can save enormous amounts of money and man power.

There are two types of self-powered AUVs to provide long-lasting and energy-efficient capabilities for surveying the deep blue. The first one is powered by the different temperatures in the water column. The SOLO-TREC is the first successful thermal recharging float and is currently studying the wakes of tropical cyclones. Current limitations in this prototype are that the primary battery lasts only 135 dives and there needs to be drastic temperature stratification for the AUV to work effectively.

The second prototype is a recently developed and tested AUV that is based on control moment gyroscope principles and collects mechanical energy from wave movement in a flywheel mount, This system promises to extend AUV mission duration indefinitely and reduce support vessel time currently required for periodical recharging and redeployment. This system could potentially be applied to any rotating platforms such as autonomous surface vessels like buoys or boats.

Besides the cost benefits, AUVs are essential in a wide range of oceanographic venues such as marine geoscience, and are increasingly used in scientific, military, commercial, and policy sectors. There contributions have advanced marine research in submarine volcanism and hydrothermal vent studies, mapping and monitoring of low-temperature fluid escape features, and mapping of other water environments. Along with the technological advancements in energy efficiency, there have been advancements in hovering, long endurance, extreme depth, and rapid response capabilities that will allow for leaps in oceanographic research

(http://digital-library.theiet.org/content/journals/10.1049/iet-rpg.2015.0210).
(http://www.sciencedirect.com/science/article/pii/S0025322714000747).
(http://ieeexplore.ieee.org/document/7485367/).

Wynn, Russell B., et al. "Autonomous Underwater Vehicles (AUVs): Their past, present and future contributions to the advancement of marine geoscience." Marine Geology 352 (2014): 451-468.

Y. Chao, "Autonomous underwater vehicles and sensors powered by ocean thermal energy," OCEANS 2016 - Shanghai, Shanghai, 2016, pp. 1-4.
doi: 10.1109/OCEANSAP.2016.7485367

Trucks and Sedans for Tesla

by Mary-Catherine Riley

Tesla unveiled its "Master Plan, Part Deux" in July of 2016 that included a Roadster convertible, a Tesla Model 3, and a plan to "expand to cover the major forms of terrestrial transport." This, by nature, included the addition of a Tesla truck and semi truck.

In April of 2017, Musk tweeted "Tesla Semi Truck unveil set for September. Team has done an amazing job. Seriously next level". The advancement makes sense in the context of the company's mission statement in "Part Deux." However, Tesla is not the only one innovating in this market. One of Tesla's cofounders, Ian Wright, runs Wrightspeed which builds hybrid battery heavy duty trucks. Nikola Motors is using hydrogen batteries and natural gas to extend the range of semi truck mileage. There is skepticism about the Tesla trucks as a current passenger car can barely break a 300-mile range; therefore, it is challenging to envision an electric truck towing tens of thousands of pounds of cargo and still achieving a decent range. However, part of the 'Master Plan' states that, "We believe that Tesla Semi will deliver a substantial reduction in the cost of cargo transport, while increasing safety and making it fun to operate". One of Tesla's goals include autonomous driving technology for semi trucks. Autonomous driving would reduce the two main costs in the trucking industry: gas and labor. Though this modernization would not occur in the first generation of semis.

A second innovation in Tesla's plan is to unveil the model of a pickup truck in 6 to 9 months and have them go into production in 3 to 4 years. Musk expressed an interest in the possibility of using a swappable battery in order to eliminate the need to wait for the car battery to recharge.

Estimations say that the Tesla Model 3, a compact, mainstream priced sedan will help boost Tesla production from 100 thousand in 2016 to 500 thousand in 2018.

It will be fascinating to see how their innovations continue to push forward renewable energy in the automotive industry.

(http://www.nbcnews.com/business/autos/tesla-building-pickup-semi-truck-n746611).
(https://electrek.co/guides/tesla-pickup-truck/).
(http://www.nbcnews.com/business/autos/tesla-building-pickup-semi-truck-n746611).
(https://electrek.co/2017/04/13/tesla-semi-all-electric-truck-september-elon-musk/).
Eisenstein, Paul A. "Tesla Is Building a Pickup and a Semi-Truck." NBCNews.com. NBC Universal News Group, 14 Apr. 2017. Web. 19 Apr. 2017.
Lambert, Fred. "Tesla Pickup Truck." Electrek. N.p., 14 Apr. 2017. Web. 19 Apr. 2017.
Lambert, Fred. "Tesla Semi All-electric Truck to Be Unveiled in September and Be 'next Level', Says Elon Musk." Electrek. N.p., 13 Apr. 2017. Web. 19 Apr. 2017.

2017 Brings More Tesla Superchargers and New Usage Regulations

by Bianca Rodriguez

Tesla Superchargers are currently the best and fastest charging option for long-distance travelers driving one of Tesla's all-electric vehicles. A Supercharger takes a mere 30 minutes to replenish batteries from 10% to 80% charge, enough time to take a restroom break or grab a coffee during a long trip; or 75 minutes to reach a full 100% charge, enough time for a meal at a nearby restaurant. A battery charged at 80% will provide about 170 miles of driving range, which should be enough to reach the next Supercharger along some of the more popular routes. Even so, Tesla is continuing to increase the number of Supercharger locations around the world to fill the need of an increasing population of Tesla drivers. This is especially necessary due to the new Tesla Model 3, which is expected to be available after 2018. Starting at $35,000, the Model 3 is Tesla's most affordable car and will most likely increase the number of Tesla drivers as more people will be able to afford these high-tech full electric vehicles.

Access and usage of Superchargers have been free to Model X and Model S owners since their introduction. However, new Tesla owners who have ordered their vehicles after January 15, 2017 will receive a limited amount (about 1,000 miles worth) of charging per year, followed by fees for any additional charging. Any charging done past the allotted amount will be automatically billed to a credit card or bank account linked to the vehicle. Most Superchargers will have a fixed price rate per kWh (kilowatt hour) based on the charger's location, but some Superchargers will be priced per minute based on the selected charging speed. Additionally, Tesla announced a new solution to overcrowding at charger locations. An "idle fee" has been implemented for cars left plugged in to a Supercharger after reaching full charge, encouraging drivers to move their vehicles upon charge completion. Fortunately, these fees will only be applied at crowded charging stations where at least half the chargers are in use.

https://www.tesla.com/supercharger
https://www.tesla.com/support/supercharging
https://www.tesla.com/support/supercharger-idle-fee
https://www.tesla.com/model3

The 2017 Chevy Bolt and the Future of Electric Vehicles in the United States

by Bianca Rodriguez

Chevrolet, a division of General Motors, has released the first full-sized electric vehicle that is not only affordable, but practical as well. The 2017 Chevy Bolt, (not to be confused with the Chevy *Volt* hybrid) has won numerous awards including 2017 Motor Trend Car of the Year and 2017 Green Car of the Year by the Green Car Journal. Starting at

$29,995 after federal tax credit, the Bolt is targeted for the average consumer, rather than just the wealthy electric car enthusiast. However, the low price does not mean compromising comfort and range, as the Bolt is a full sized 4-door/5-passenger hatchback with an advertised 238-mile range. This means that the average commuter can go several days between charges. Drivers who wish to go long-distance, on the other hand, will have to find designated charging stations along their route that can provide up to 90 miles for 30 minutes of charge time.

Compared to the Tesla Model S, which has a battery range between 200 to 300 miles, the Chevy Bolt's 238-mile range is great considering it costs only half as much. Yet, battery range is a concern that many people have when deciding to opt for a full-electric vehicle. This is a concern that Ford is hoping to address by introducing a 300-miles range electric compact SUV by the year 2020. Even with this goal, Ford believes that hybrids will be the predominant consumer choice in the near future as drivers do not have to worry about range. Another American electric car company, Lucid Motors, has recently announced a battery upgrade with a range of 400 miles for their newly announced vehicle, the Lucid Air. Expected to be available in 2018, the Lucid Air will cost upwards of $100,000. Regardless of the currently expensive price tag of larger-range electric cars, it is clear that battery technology is being continuously improved upon. What Chevy, and other U.S. car companies have shown the world is that the future of electric cars is practical and very much accessible.

http://www.chevrolet.com/bolt-ev-electric-vehicle.html
http://www.theverge.com/2016/9/13/12899752/chevy-bolt-driving-impressions-review
http://www.usatoday.com/story/tech/news/2017/01/03/ford-have-300-mile-range-ev-2020/96106868/
https://www.engadget.com/2016/12/14/lucid-motors-unveils-its-400-mile-range-luxury-ev/

The Need for Efficient Battery Recycling in a New Electric Car Era

by Bianca Rodriguez

As some of the first popular electric cars approach their 10th year on the road, their batteries are also reaching the end of their lifespan. But even when batteries have become unusable for electric vehicles on the road, they still retain a significant amount of battery capacity. Up until recently, battery recycling was an inefficient and wasteful practice as only the metals (mainly nickel, cobalt and iron) are extracted and the rest is disposed. Today, new technology is being developed to allow for greater efficiency in recycling of more of the materials in batteries. Battery Resourcers LLC is a company out of the Worcester Polytechnic Institute (WPI) in Massachusetts that has developed a method for extracting battery cathode material to be used for generating new battery cells. In particular, they focus on lithium batteries, as they are currently the most often used batteries in electric cars. As electric vehicles become more popular so does the demand for lithium, therefor being able to recycling lithium material would be beneficial. Knowing the expanding need for recycling electric vehicle batteries, Battery Resourcers are focusing their efforts in this market as they believe it will have the biggest environmental impact. Even Tesla Motors is realizing the need for lithium battery recycling. Their new Gigafactory in Nevada will host a battery recycling center on site to extract materials for new battery cells, though the technology they will be using has not been revealed.

Until better recycling becomes a streamline process, there will still be a great number of electric vehicle batteries to deal with. Battery repurposing is not a new idea, but with increased accessibility to solar technology and a more environmentally-conscious population, the idea of using a car battery to store electricity to power a home is no longer farfetched. For example, old electric-car batteries can be set up to charge in the day via solar panels or during off-peak hours, then release the electricity as needed at night or during off-peak hours. There are numerous other applications for repurposed electric vehicle batteries, but the main concern today is their safety. As the technology and standards for testing the safety of each used battery is being developed, a market infrastructure must also be set up to allow electric vehicle owners to sell their used batteries.

http://www.telegram.com/news/20170205/extended-life-wpi-trio-starts-business-recycling-lithium-ion-batteries

https://www.batteryresourcers.com

https://cleantechnica.com/2015/07/23/electric-vehicle-battery-can-recycled/

https://chargedevs.com/newswire/gigafactory-features-renewable-energy-recycling-robots-and-horses/

http://www.altenergymag.com/article/2016/11/repurposing-electric-vehicle-batteries/24964

https://cleantechnica.com/2016/10/31/repurposing-ev-batteries-underwriters-laboratory-interview/

Vantage Power: Retrofitting Diesel Buses as Hybrids

by Chloe Soltis

In 2011, Alexander Schey and Toby Schulz founded Vantage Power, a London energy start-up that manufactures hybrid powertrains that can be used in new buses or as a retrofit solution in buses that are already operating. In cities, diesel buses are often one of the most popular modes of transportation but are also one of the largest contributors to pollution. However, bus operators are often hesitant to switch their diesel fleets to hybrid due to the high costs; a hybrid bus is usually 50 percent more expensive than a diesel bus (Autodesk). Vantage Power's retrofit hybrid powertrain is a great solution to this issue since its cost is significantly less than an entire new hybrid bus. The hybrid powertrain also lowers fuel consumption by 40 percent, which allows bus operators to adhere to government environmental regulations while decreasing overhead costs. In addition, Vantage

Power's retrofit powertrains come with a technology suite that allows operators to more closely monitor their buses. The new technology is especially helpful for diagnosing and performing preventative maintenance so that the hybrid buses will be more reliable (Cooperative Energy).

A key feature of this hybrid powertrain system is its lithium ion battery. It is three times smaller in volume than other batteries in its class (Vantage Power). Designing a battery that could fit into an existing engine bay but was powerful enough to help power a hybrid bus was a challenge for Vantage Power. However, the company credits engineering a new liquid thermal management system and innovative structural integration techniques for the battery's small size. Another positive feature of Vantage Power's powertrain is that it does not take an extended period of time to install. The process of replacing the diesel runner takes approximately one week and can be done by the bus operators themselves.

In the next three years, Vantage Power hopes to expand their presence in the UK retrofit market and eventually manufacture complete hybrid vehicles.

Green Tech Heroes: Vantage Power. 2016. Cooperative Energies. Feb 5, 2017:
https://www.cooperativeenergy.coop/news-and-views/green-tech-heroes-vantage-power/
Vantage Power. http://vantage-power.com.
Vantage Power Named 2015 Inventor of the Year. 2016. Autodesk. Feb 5, 2017:
http://blogs.autodesk.com/inthefold/vantage-power-named-2015-inventor-of-the-year/.

Move Systems: Bringing Clean Energy to the Mobile Food Industry

by Chloe Soltis

Move Systems is a New York start-up that develops and manufactures products that reduce energy waste in the mobile food industry. Food carts operate on the street, where there is no direct supply of electricity. Therefore, these vendors run gas generators to operate their businesses; however, this can cost $5000-$6000 a year for a normal food cart (Tech Crunch). Generators are also a main source of noise and air pollution on city streets. In 2013, it was estimated that New York City's 3,000 mobile food carts' pollution emissions were comparable to that of 10,000 cars (Move Systems). Fortunately, Move Systems has developed two products to help combat these issues and make mobile food vendors more environmentally friendly.

Move Systems first developed the Simply Grid, an electricity pedestal that can be installed anywhere in a city where mobile food vendors are stationed. Each pedestal has outlets that can connect to the vendor's cart and will quietly power the business. The pedestal is metered and is also connected to Simply Grid's cloud-based technology system for billing and accounting purposes. Therefore, a vendor can plug her vehicle into a pedestal and wirelessly connect to the Simply Grid application with her smartphone to start and end the flow of electricity. Move Systems is also hoping that Simply Grid pedestals will be used to power other city vehicles that idle for extended periods of time such as ambulances.

Move Systems has also created the MRV100, or a hybrid food cart. The cart runs on a combination of compressed natural gas and solar panels and can be connected to an electrical outlet if one is available (CityLab). In comparison to a generated-power cart, the MRV100 is dramatically quieter and has high-quality refrigeration equipment, which prioritizes the cart's food safety. The hybrid also has the potential to be programmed for credit-card transactions, pre-ordering through smartphone applications, and electronic inventory (CityLab). While Move Systems has designed and tested these two products in New York City, it believes that they can be successfully implemented in any city.

Etherington, Darrell. Simply Grid Is Putting Clean, Easy Power Where People Need It. Tech Crunch. Feb 28, 2017: https://techcrunch.com/2014/05/05/simply-grid-is-putting-clean-easy-power-where-people-need-it/.

Goodyear, Sarah. Is This the Mobile Food Cart of the Future? CityLab. Feb 28, 2017: http://www.citylab.com/design/2015/05/is-this-the-mobile-food-cart-of-the-future/393053/.

Move Systems. https://www.movesystems.com.

XStream Trucking: GapGorilla Makes Trucks More Fuel-Efficient

by Chloe Soltis

The trucking industry is one of the driving forces of the American economy. However, truck drivers make up only 4% of the traffic on the road, yet the industry spends nearly 100 billion dollars on fuel every year and its emissions make up 26.6% of the United States' total carbon emissions (XStream). XStream Trucking saw the need to make trucks more fuel-efficient and developed GapGorilla.

When the XStream team began researching improvement methods, they realized that the gap that exists between the trucking cab and trailer causes a substantial amount of drag. The gap is important for helping the truck turn at slower speeds yet serves no purpose when the truck is barreling down the highway. Therefore, GapGorilla is a device that is attached to the cab of the truck and extends its panels to bridge the gap between the trailer and cab when the truck is traveling at speeds higher than 40 miles per hour. It closes at speeds lower than 40 miles per hour to allow the truck to properly turn. Eliminating this drag allows the truck to consume up to 4% less fuel (XStream). In addition, the GapGorilla is considered lightweight at 150 pounds and is automatic.

In 2016, XStream Trucking successfully tested their product both on virtual computer models and on physical highways (FLOW). GapGorilla also won multiple awards including second place at 2016

Department of Energy National Cleantech Up competition. Finally, on June 24, 2016, the company signed its first contract worth nearly $600,000 to install GapGorillas on a fleet of trucks. XStream Trucking is now talking with other customers including Pepsico on installing more GapGorillas in the future.

"The First Look West (FLoW) 2016 Winners." FLOW. Caltech. April 11, 2017: http://www.flow.caltech.edu/2016-winners.
XStream Trucking. http://www.xstreamtrucking.com.

About the Authors

The authors of this book are students at the Claremont Colleges. The book is a work product of Biology 137: Environment, Economics, and Policy Clinic taught by Emil Morhardt in the W.M. Keck Science Department of Claremont McKenna, Pitzer, and Scripps Colleges.

The students' task was to write journalistic summaries of descriptions of interesting new innovations in energy culled from whatever sources they wished, capturing the essence of the innovations but eschewing technical terms to the extent possible—to become, in effect, science writers. The summaries were due weekly and were returned with editorial comments shortly thereafter.

The editor is Roberts Professor of Environmental Biology at Claremont McKenna, Pitzer, and Scripps colleges. He remembers how difficult it is to learn to write and appreciates the professionalism shown by these students.